THE DANCE OF HOMO STULTUS

Matteo Preabianca

The story, all names, characters, and incidents portrayed in this book are fictitious. No identification with actual persons (living or deceased), places, buildings, and products is intended or should be inferred.

Paper back ISBN-13: 978-1-963272-98-7
Hardcover ISBN-13: 978-1-963272-99-4
Dyslexic Friendly ISBN-13: 978-1-963272-97-0

ShelteringTree.Earth, LLC Publishing
PO Box 973, Eagle Lake, FL 33839

Did you enjoy this book?

We love to hear from our readers.

Please visit the author and illustrator at
ShelteringTreeMedia.com

IN MEMORIA

To all the animals who have lost their lives due to human actions—
May their stories inspire us to choose kindness, wisdom, and respect for all living beings.

CONTENTS

INTRODUCTION

In the vast and complex landscape of life on Earth, we often find ourselves comparing the differences and similarities between humans and animals. While we can boast a wide range of cognitive abilities and a sophisticated capacity for transmission, we cannot ignore the fact that we share this planet with countless other forms of life, each with its own unique traits.

In the pages that follow, we will embark on an intriguing and often sarcastic journey through the many facets that define the relationship between humans and the rest of the animal kingdom. With a lighthearted lens, we'll explore topics ranging from boredom to longevity, from forgiveness to habitat destruction, aiming to shed new light on the dynamics that connect us with the other inhabitants of our world.

In this book, we'll adopt an unusual perspective: one that favors the animals. We'll examine their unique features, their fascinating habits, and their often-overlooked viewpoints. Through this provocative approach, we hope to challenge our preconceived notions and promote greater awareness and compassion for the other forms of life with whom we share the world.

Get ready for a sarcastic yet enlightening journey into the world of animals and their relationships with humans. Whether you're a passionate animal rights advocate or simply curious, this book invites you to explore the boundaries of intellect and empathy in a context rich with irony and wisdom.

1 ORIENTATION

Cats, our loyal companions, exhibit an exceptional talent for spatial orientation that rivals our most sophisticated GPS (Global Positioning Service) technology. While we humans struggle to find our way using paper maps, apps, and occasional requests for directions, our feline friends effortlessly navigate the world around them, seemingly guided by a finely tuned internal GPS system attuned to their surroundings. Forget relying on Siri or Google Maps; cats have an innate ability to find their way home even after the most complex adventures. Picture this: a lost cat and a lost human wandering through the same neighborhood. The human, clutching a smartphone and frowning in frustration, aimlessly walks around, hoping for a miracle from the satellites above. Meanwhile, the cat moves confidently, occasionally pausing to inspect a bush or chase a butterfly that flutters by, always maintaining an impeccable sense of direction.

Underwater, cetaceans showcase their remarkable navigation skills, weaving effortlessly through ocean currents with a precision that puts human navigation technology to shame. These intelligent marine mammals possess an intricate understanding of echolocation, using sound waves to map their environment and locate targets with astounding accuracy. Imagine a pod of dolphins watching humans struggle with underwater navigation instruments, shaking their heads amusedly at our clumsy attempts to navigate their domain. While we rely on bulky sonar systems and GPS trackers, these graceful cetaceans glide effortlessly through ocean depths, guided by their natural sonar, navigating even the most perilous waters without trouble.

In the insect realm, we find the humble worker bee, a tiny navigator with a talent for precision that rivals any human-made device. Through a complex dance known as the "waggle dance," worker bees communicate precise directions to food sources located miles away from their hive. This remarkable form of contact allows worker bee colonies to efficiently forage for nectar and pollen, ensuring the entire hive's survival. Imagine a team of scientists marveling at the worker bee's intricate connection system, trying to replicate its efficiency in our technologies. While we develop complex algorithms and link protocols, worker bees continue to rely on their innate ability to convey spatial information through a series of elaborate movements and vibrations.

From the skies to the ocean depths, from the tiniest insects to the most majestic marine mammals, the animal kingdom abounds with creatures possessing extraordinary navigation abilities. As humans, we often take pride in our technological achievements and intelligence, but we should not overlook the talents surrounding us among our fellow earthlings. In a world so heavily reliant on gadgets and gizmos, animals remind us that sometimes the most extraordinary navigation tools are the ones we are born with. So, here's to the whisker-guided explorers, the echolocation experts, the dance-driven seekers, and all the other marvels of navigation in the animal realm. May we continue to marvel at their skills and learn from their innate wisdom.

As we continue to explore the astonishing world of animal navigation, we encounter creatures whose abilities challenge our understanding of spatial orientation and push the boundaries of what we thought possible. Take the incredible monarch butterfly, for instance, a delicate yet resilient insect that undertakes epic migrations thousands of miles long, crossing continents with extraordinary precision. Despite their fragile appearance, monarch butterflies navigate vast distances with astonishing accuracy, guided by a combination of innate instincts and environmental cues. Imagine witnessing the spectacle of millions of monarch butterflies embarking on their annual journey, from their wintering grounds in the mountains of Mexico to their summer breeding sites in the northern United States and Canada. These majestic insects, weighing less than a gram and with wings as delicate as tissue paper, defy the odds as they navigate diverse landscapes and changing weather conditions.

At the heart of the monarch butterfly's navigational toolkit lies their acute sensitivity to environmental signals, particularly the sun's position. They use the sun as a compass, orienting themselves based on its position in the sky. By maintaining a constant angle relative to the sun's rays, they can determine the direction of their journey, ensuring they stay on course during their long voyage. However, the sun alone is not sufficient to guide the monarchs. These extraordinary insects also possess a sophisticated internal clock, enabling them to compensate for the sun's movement throughout the day. As the sun shifts position in the sky, monarchs adjust their flight path accordingly, ensuring they stay on track even as daylight turns to dusk.

In addition to celestial signals, they rely on environmental landmarks to navigate vast distances. From mountain ranges to rivers and coastlines, monarchs use visual cues to confirm their position and make necessary adjustments to their flight route. Even in unfamiliar territories, they demonstrate an impressive ability to navigate with precision, ensuring they reach their destination despite any challenges they may encounter along the way. Perhaps most astonishing of all is their ability to pass on this navigational knowledge from one generation to the next. Each year, they embark on their epic migration without ever having made the journey before. Instead, they rely on a combination of genetic programming and environmental signals to guide them along the same migratory routes as their ancestors. This remarkable inheritance of knowledge ensures that each generation of monarch butterflies can successfully complete their journey, despite never having undertaken it themselves.

It is a testament to the persistence and adaptability of these incredible insects and a reminder of life's interconnectedness on Earth. In a world where technology reigns supreme, the humble monarch butterfly serves as a powerful reminder of the innate wisdom encoded in the natural world. While we boast of our scientific and technological advancements, there are still mysteries waiting to be discovered in the world around us. By studying the remarkable navigational abilities of creatures like these butterflies, we can gain valuable insights into the complex dynamics of the natural world and our place within it.

The monarch butterfly's migration is a testament to the wonders of nature. As they undertake this incredible feat, they face numerous challenges along the way. From harsh weather conditions to natural predators, the journey is fraught with danger. Yet, despite these obstacles, the butterflies persist, driven by an innate urge to complete their migration and ensure future generations' survival. One of the greatest challenges they face during their migration is the vast distance they must cover. Spanning thousands of miles, the monarchs must navigate through diverse landscapes, from sprawling deserts to densely populated urban areas. Urban development, agricultural expansion, and habitat loss have all impacted their populations, reducing the availability of crucial resources like nectar-rich flowers and milkweed plants, on which monarchs depend for food and reproduction.

Climate change also poses a significant threat to their migratory journey. With rising temperatures and increased unpredictability in weather patterns, they must contend with changing environmental conditions that can impact their ability to navigate and survive. Extreme weather events, such as hurricanes and droughts, can disrupt the monarchs' migratory patterns and destroy critical habitats, further endangering these magnificent insects. Despite these challenges, monarch butterflies continue to persevere, aided by conservationists and researchers working tirelessly to protect and preserve their habitats. Organizations like the Monarch Joint Venture and the Xerces Society for Invertebrate Conservation are actively engaged in efforts to restore monarch habitats, promote milkweed conservation, and raise awareness about the importance of protecting these butterflies and their migration.

Citizen scientists also play a crucial role in monarch conservation efforts, contributing valuable data and observations that help researchers monitor monarch populations and understand their migratory behaviour. Through initiatives like Monarch Watch and Journey North, people of all ages can participate in monarch monitoring programs and contribute to scientific research aimed at conserving these iconic insects. Perhaps the most inspiring aspect of their migratory journey is the sense of wonder and awe it inspires in people worldwide. Each year, millions of monarch butterflies captivate the public with their breathtaking beauty and extraordinary perseverance as they embark on their epic migration. From schoolchildren to seasoned researchers, people from all walks of life are drawn to the sight of monarchs in flight, marveling at their incredible journey and the mysteries of the natural world.

In a world increasingly interconnected and driven by technology, the monarch butterfly's migratory journey serves as a poignant reminder of the importance of conserving and protecting our planet's biodiversity. By preserving habitats and supporting efforts to safeguard species, we can help ensure a brighter future for generations to come. Working together to conserve their habitats and protect their populations, we can help ensure that monarch butterflies continue to inspire and captivate us for generations to come.

2 THE BONOBO EXPERIMENT

In the lush, vibrant heart of the Congo Basin, amid dense vegetation, lies a captivating experiment within nature's own laboratory—the world of the bonobos. These enigmatic primates have long been subjects of scientific investigation, particularly to understand their social dynamics and, even more intriguingly, their approach to intimacy.(Call, Kuta, 2024)[1] As we dive into this realm,

---

[1]Call, J. (2024, September 3). Bonobos Can Tell When a Human Doesn't Know Something. *Scientific American*. Retrieved November 28, 2025, from https://www.scientificamerican.com/article/bonobos-can-tell-when-a-human-doesnt-know-something/
Kuta, S. (2024, September 5). When Bonobos Know What You Don't, They'll Tell You. It's a Sign of a Cognitive Ability Called 'Theory of Mind'. *Smithsonian Magazine*. Retrieved November 28, 2025, from https://www.smithsonianmag.com/smart-news/when-bonobos-know-what-you-dont-theyll-tell-you-its-a-sign-of-a-cognitive-ability-called-theory-of-mind-180985991/

we embark on a journey that challenges preconceived notions about the boundaries between humans and animals.

The bonobo, often dubbed the "hippie ape" or the "make love, not war" primate, is closely related to the chimpanzee. Unlike their more aggressive relatives, bonobos have created a society founded on cooperation, transmission, and, yes, an unconventional approach to intimate relationships.

In the bonobo community, sex is not just a means of reproduction; it's a cornerstone of social bonding. From casual encounters to complex social negotiations, bonobos engage in sexual activities with a freedom that might make even the most liberal humans blush. This stands in stark contrast to the behaviour of chimpanzees, where conflicts are often resolved through aggression and violence.

Picture this: a group of bonobos facing a conflict over a particularly coveted food source. What follows is not a scene of aggression and fighting but rather an intricate dance of desire. Bonobos use sexual activity to diffuse tension, forge alliances, and build social connections. It's their version of conflict resolution, a strategy that transcends the primal aggression observed in many other species.

Yet, can we not find parallels in our own human interactions? In the sophisticated landscapes of our societies, the dance of desire is a ubiquitous rhythm, played out in the nuances of human relationships. While we may not engage in public displays like our primate cousins, the undercurrents of flirtation, attraction, and the delicate choreography of human courtship are woven into the fabric of our daily lives.

In the world of bonobos, love is not confined to monogamous pairs or traditional family structures. Instead, it's a fluid and dynamic force woven into their social fabric. Bonobos engage in a variety of sexual behaviours, from simple genital touching to full intercourse, often regardless of age or gender. The bonds forged through these encounters act as the glue that holds their society together. Unlike the often turbulent relationships seen in human communities, bonobos seem to navigate the complexities of intimacy with an ease that would astonish any couples therapist.

However, when we turn our gaze inward, can't we see parallels in our own diverse approaches to love? Even human societies have evolved to recognize a wide range of relationships—from conventional to innovative, from monogamous to polyamorous. The human capacity to form complex emotional bonds spans a vast spectrum, reflecting the complexity of our species' approach to love and intimacy.

As we contemplate the bonobo experiment, it's impossible to ignore the stark contrast between their conflict resolution approach and ours. While we wield weapons and wage wars, bonobos extend a hand—sometimes quite literally—to embrace peace through pleasure. Violence is an alien concept in bonobo society. Their unique approach to intimacy, characterized by empathy and cooperation, serves as a powerful reminder that perhaps there's much more we could learn from our genetic relatives than we're willing to admit.

Moreover, let's not overlook the lessons we could draw from bonobos in our own struggles with violence. Can we not imagine a world where, like the bonobos, we channel our primal impulses toward cooperation and understanding rather than aggression? The bonobo experiment challenges us to reconsider our approach to conflict resolution and the potential of empathy as a guiding force in human society.

In the grand tapestry of life, this "experiment" stands as a testament to the vast spectrum of possibilities that exist in the animal kingdom. It challenges us to rethink our assumptions about the boundaries between humans and animals, pushing us to appreciate life's diversity and the myriad of ways species navigate the complexities of existence.

So, as we marvel at the bonobos, let's take a moment to reflect on the lessons they offer. In their intricate dance of desire, their bonds of love, and their rejection of violence, perhaps we can find inspiration to build a world where empathy and cooperation guide our interactions, just like our fascinating primate cousins in the heart of the Democratic Republic of the Congo(DRC). Through understanding and embracing the wisdom of our adaptive relatives, we may discover new ways to coexist more harmoniously with one another and the natural world.

As we delve deeper into the fascinating world of bonobos and their social dynamics, it's imperative to broaden our scope and draw comparisons to other primates, particularly our closest relatives, the chimpanzees.

Chimpanzees, like bonobos, belong to the Pan genus and share about 98.8% of their DNA with humans (The Chimpanzee Sequencing and Analysis Consortium, 2005)[2]. Despite this genetic closeness, their social structures and behaviours can be surprisingly different. While bonobos are characterized by their peaceful and egalitarian societies, chimpanzee communities often exhibit hierarchical dominance and aggression, especially among males.

In chimpanzee societies, conflicts over resources like food or territory are typically resolved through displays of power, physical aggression, and even intergroup warfare. Male chimpanzees form alliances and compete for dominance within a strict social hierarchy, often resorting to violence to maintain or improve their status.

This contrasts sharply with the bonobo approach, where conflicts are mitigated through sexual behaviours and social bonding rather than aggression. In bonobo communities, both males and females engage in sexual activities indiscriminately, fostering a sense of cooperation and solidarity that transcends individual differences.

The differences between bonobos and chimpanzees provide valuable insights into the various social strategies adopted by primates. While chimpanzees emphasize dominance and

---

[2] The Chimpanzee Sequencing and Analysis Consortium. (2005). Initial sequence of the chimpanzee genome and comparison with the human genome. *Nature*, *437*(7055), 69–87. https://doi.org/10.1038/nature04072

competition, bonobos prioritize cooperation and conflict resolution through affectionate behaviours such as grooming and sexual interactions.

By comparing the social dynamics of bonobos and chimpanzees, we gain a deeper understanding of the Darwinian origins of human behaviour. The divergent paths taken by these closely related species highlight the complex interplay between genetic predispositions and environmental factors in shaping social structures and behaviours.

Through interdisciplinary research that integrates perspectives from primatology, anthropology, psychology, and sociology, we can continue to uncover the underlying principles governing social dynamics across species and identify strategies to promote peace and cooperation in human societies.

Love, an emotion often associated with humans, transcends species boundaries. Across the animal kingdom, displays of affection, bonding, and care can be observed, illuminating the intricate web of emotions that permeate the natural world. Among the many species that exhibit remarkable displays of love, the bonobo stands out for its unique social structure and intimate interactions. This chapter delves into the realm of animal affection, focusing particularly on the fascinating parallels and distinctions observed in the love lives of animals, with a special emphasis on bonobos.

Animal love is a multifaceted phenomenon, encompassing various forms of social bonding, care, and affection. From the tender grooming sessions of primates to the intricate courtship dances of birds, examples abound in the animal kingdom. One of the most evident demonstrations of love is maternal care, evident in a wide range of animals. Mammals, in particular, display deep maternal instincts, nurturing their young with devoted commitment.

Among social mammals like elephants and some marine mammals, familial bonds are strong, with individuals showing empathy, cooperation, and altruism toward one another. Wolves, for instance, form close-knit family units where pack members display loyalty and affection for each other. Similarly, in primate societies, social grooming serves not just hygienic purposes but also promotes social cohesion and emotional connections among group members.

In conclusion, animal love encompasses a wide range of behaviours and emotions, reflecting the intricacies of social bonds and relationships within the animal kingdom. From maternal care to social grooming, animals exhibit various forms of affection that contribute to the cohesion and stability of their communities. Among these, bonobos stand out for their unique socio-sexual behaviours, empathy, and peaceful social structure.

Through their liberal approach to sexuality, bonobos offer valuable insights into the speciation origins of love and cooperation in humans. By studying the love lives of animals, we gain a deeper understanding of the complex interplay between emotions, social dynamics, and progressive pressures that shape the natural world. Unraveling the mysteries of animal affection, we ultimately come to understand the profound universality of love across species.

3 BOREDOM

Ah, boredom, that exclusive state of mind where time slows down to a snail's pace, and the mundane becomes captivating. Humans often complain about their encounters with boredom, seeing it as a curse upon their existence. But let us cast our gaze upon the animal kingdom, where boredom reigns supreme with an air of nonchalance and superiority. Picture the majestic sloth, lazily sprawled on a branch, eyes half-closed in blissful indifference. To the human observer, the sloth may appear sluggish, unmotivated, and, dare I say, bored. But oh, how wrong we are!

The sloth is not bored; it is savoring the simple pleasures of existence, unburdened by the frenetic pace of human life. Its serene lifestyle is a testament to the art of unhurried living. Contrast this with the plight of modern humans, trapped in a relentless cycle of productivity and stimulation. We fill our days with endless tasks and entertainment, desperately trying to ward off the dreaded specter of boredom. Yet, in our frantic pursuit of novelty, we overlook the deep wisdom of our fellow animals. Consider the noble tortoise, whose slow and deliberate movements hide a profound inner peace.

While humans scurry about frantically, the tortoise glides serenely, wholly indifferent to the passage of time. To the tortoise, boredom does not exist, only the quiet contentment of life lived at its own pace. And what about the domestic cat, that paragon of feline indifference? With an air of regal disdain, it surveys its domain, each movement a testament to the art of doing nothing. While humans endlessly scroll through streams of digital content, the cat lounges in the sun, its only concern the occasional pursuit of a wandering speck of dust. Perhaps the most enviable creature in the realm of boredom is the humble housefly. With its brief lifespan and limited cognitive faculties, the fly knows nothing of boredom. It flits from place to place, its tiny brain untroubled by the existential dread that plagues its human counterparts.

For the fly, every moment is a marvel, every speck of dust a source of infinite fascination. Let us not forget the wisdom of the humble cow, contentedly chewing its cud in the pasture with an air of serene satisfaction. While humans rush from one activity to another, the cow grazes calmly, embodying the essence of tranquility. Or consider the little goldfish, swimming in the pond in a perpetual dance of grace and simplicity. While humans crave constant stimulation, the goldfish finds comfort in the rhythmic patterns of its existence. And who could overlook the earthworm, burrowing through the soil with a quiet determination that belies its lack of sophistication? While humans struggle to find meaning and purpose, the earthworm embraces its role in the ecosystem with silent humility. In the grand tapestry of existence, boredom is but a thread, weaving itself into the fabric of life.

And while humans may struggle to come to terms with this fundamental aspect of our shared experience, our fellow animals have long made peace with it. So let us shake off the chains of productivity and frantic activity and learn to embrace the stillness of boredom, as our animal brethren have for countless generations. In the end, it is in moments of quiet and tranquility that we truly find ourselves.

Continuing our exploration of boredom in the animal kingdom, let us now turn our attention to the fascinating world of insects, where creatures of all shapes and sizes navigate the intricacies of monotony with remarkable ingenuity. Take, for example, the industrious bee, tirelessly flying from flower to flower in search of nectar. While humans might consider its repetitive task mundane, the bee finds fulfillment in the rhythm of its work, contributing to the hive's greater collective with unwavering dedication. Similarly, the diligent ant colony buzzes with activity as its inhabitants work ceaselessly to secure food and shelter. While each ant might perform seemingly repetitive tasks, together they form a complex and highly organized society, where every action serves a purpose in the grand scheme of the colony's life.

In the realm of arachnids, the patient spider meticulously weaves its delicate web, waiting for its next meal to be ensnared in its intricate trap. While humans may grow restless in moments of stillness, the spider understands the power of patience, knowing that success often comes to those who wait. Yet it is not only the creatures of the land that offer insights into the nature of boredom. Beneath the surface of the water, the curious clam burrows into the sand, its slow and deliberate movements a testament to the art of patience. While humans may crave constant stimulation, the clam finds satisfaction in the simple act of existence, rooted in the tranquil depths of the ocean floor. And what of the mighty whale, effortlessly gliding through the vast expanse of the ocean? While humans may fear the vastness of the open sea, the whale embraces it, finding comfort in the endless horizon and the gentle rhythm of the waves. Even in the realm of reptiles, creatures like the ancient tortoise and the serene sea turtle remind us of the beauty of a life lived at a leisurely pace. While humans rush through life in a constant state of frenzy, these wise reptiles understand the value of taking things slow, savoring each moment as it comes. And let us not forget the countless species of birds that grace the skies with their graceful flight.

From the majestic eagle to the playful hummingbird, each avian creature offers its perspective on the art of boredom, finding comfort in the freedom of the open air and the endless possibilities of the world around them. In every corner of the natural world, from the smallest insect to the largest mammal, creatures of all kinds offer us valuable lessons on the nature of boredom. They remind us that boredom is not a curse to be avoided but a natural part of the human experience—an opportunity to slow down, reconnect with ourselves, and find joy in the simple pleasures of existence. So let us heed the wisdom of our fellow animals and learn to embrace the beauty of boredom, knowing that in those moments of stillness and quiet, we may discover the true richness and depth of life itself.

4 HOW MUCH WE TALK

Ah, the age-old debate about language and contact. Humans, with their vast vocabulary and linguistic prowess, often proclaim themselves superior in the realm of expression. But let's take a moment to admire the subtle sophistication of our earthly companions, the animals, unfairly relegated to the margins of this linguistic spectacle.

Humans, with their inflated sense of linguistic greatness, boast a vast repertoire of words. They use language to convey complex ideas, share stories, and express emotions. Oh, the glory of human conversation, where verbosity reigns supreme!

One might argue that animals lack the intricate web of words humans weave, but let us not forget the charm of simplicity. Animals communicate efficiently and directly, without the need for an excess of words. No lengthy monologues about the weather, politics, or existential musings. Animals get straight to the point.

While humans might pride themselves on their linguistic diversity, animals offer a symphony of connection that transcends mere words. Birds sing intricate melodies, whales engage in the art of underwater arias, and crickets serenade the night with their harmonious rhythms. Who needs words when one can communicate through the universal language of music?

Humans may have composed sonnets and operas, but animals, in their humble simplicity, create sound masterpieces that resonate with the heartbeat of nature itself. Perhaps it is time for Homo Stultus to trade their verbose verses for a more melodic approach to link.

Ah, the ballet of body language, where animals gracefully dance through life with gestures that convey more than a thousand words. Humans may rely on facial expressions and hand movements, but animals have perfected the art of non-verbal transmission with elegance.

From the majestic fan of a peacock displaying its iridescent plumage to the subtle flick of a cat's tail, animals communicate with a finesse that outshines human pantomime. No need for linguistic acrobatics; a simple wag of the tail or a raised eyebrow speaks volumes in the animal kingdom.

In their quest for linguistic dominance, humans often overlook the profound wisdom that comes from silence. Animals, masters of the art of silent contemplation, understand the power of unspoken connection. There is a certain eloquence in the stillness of a wise owl or the calm gaze of a sea turtle.

Consider the sly grin of a fox as it eludes its pursuers or the mocking call of a parrot mimicking human language. Animals, with their subtle gestures and clever imitations, excel in the fine art of sarcasm. While humans wield words like swords, animals wield sarcasm as a secret language, leaving us to ponder the depth of their wit.

As we reflect on the seemingly insurmountable linguistic gap between humans and animals, we should not dismiss the exquisite eloquence of our terrestrial companions.

If humans are the poets of spoken word, animals are the lyrical geniuses of the non-verbal realm. Take, for instance, the hypnotic dance of fireflies on a summer night, each spark telling a tale of romance and courtship. Humans may craft verses that move the heart, but animals, in their silent poetry, paint pictures with the brushstrokes of their existence.

Consider the delicate brushstrokes of a butterfly's flight or the rhythmic undulation of a snake through tall grass. Animals, without the need for syllables and consonants, create poetry in motion. Each creature contributes a verse to the grand epic of nature, where the language of movement surpasses the limits of spoken expression.

In the human world, subtlety often takes a backseat to the clamor of verbosity. Animals, however, have perfected the art of saying more with less. Take, for example, a cat's purr—a subtle vibration that conveys contentment, trust, and a touch of inscrutability. The slight ear twitch of a dog or the subtle eyebrow lift in the animal realm speaks volumes without the need for a vocabulary as large as a dictionary.

Humans delight in creating metaphors and allegories to convey deeper meanings. Yet animals have woven allegorical tales since time immemorial, carving their narratives into the very fabric of folklore and mythology.

Consider the wisdom of the tortoise in Aesop's fables or the mythical phoenix rising from the ashes. Animals, in their splendid symbolism, teach profound lessons through the allegorical mosaic of their existence. While humans may boast of their storytelling prowess, animals, through the ages, have been the unsung bards of allegory, imparting timeless truths with a simple nod or flick of the tail. Humans often pride themselves on the depth and complexity of their interpersonal relationships, yet animals engage in a dance of connection that rivals the most intricate human bonds. Take, for example, the collaborative efforts of ants in building elaborate colonies or the synchronized swimming of delphinids in the open sea. While humans navigate the maze of relationships with words, animals create symphonies of connection through shared endeavors.

The cooperative feats of animals go beyond mere survival; they resonate with the fundamental rhythm of harmony that echoes throughout the natural world. From the social complexities of elephants to the communal roosting of bats, animals demonstrate that the language of cooperation is spoken in the universal tongue of unity. Perhaps it is time for Homo Stultus to set aside their self-proclaimed crown of social competence and humbly learn from the orchestral brilliance of their fellow inhabitants.

Humans often engage in elaborate rituals and ceremonies to mark significant events, from weddings to funerals. Yet animals, with their innate understanding of rhythm and tradition, partake in rituals that transcend the confines of human formality.

Humans often boast about the complexity of their language as a form of contact. It's a set of verbal and non-verbal conventions used to express ideas and desires. We use words when we speak to convey our needs, and we cry or make faces when we want to express feelings. Animals, or rather Non-Humans, show other signs of link, like a dog wagging its tail when happy or a bird singing a song to attract a mate. But do animals have a language of their own? Scientists are still uncertain about the answer.

Researchers claim that Non-Human Animals do not have a true language like humans. However, they communicate with one another through sounds and gestures. Animals have a set of innate qualities they use to signal their feelings, but these are not similar to the formed words we see in human language. Human infants show these same forms of transmission from birth when they cry and gesture. But they slowly learn the words of language and use them as a form of contact.

If human infants were separated from other humans at birth, they would not learn the words of language and would be unable to communicate with other humans. They would resort to sounds and gestures as their primary form of connection. However, in the animal kingdom, if animals grow up on their own from birth, they still behave and communicate in the same way as others of their species. And what about animals like dogs that understand commands or birds that can "talk"? Dogs can be trained to follow certain commands like "sit," "come," and "roll over," but does that mean they understand language and can use it too? Dogs are known for being skilled at reading their owners' intentions and do not respond to the actual words but to the tone in which they are spoken. So, if you say "bad dog" in a cheerful tone, the dog will wag its tail. If you say "good dog" in a harsh tone, the dog will tuck its tail between its legs. Birds in captivity are known to be able to "talk"—it's believed that for them, it doesn't mean anything, and they are simply imitating the sounds they hear.

There is no doubt that animals communicate with each other in one way or another in response to different stimuli like hunger or fear. Human language is creative and consists of unique features that give us the ability to engage in abstract and analytical ways. This doesn't mean we are smarter; I'd rather say we are more complicated than complex. The fact that we have created imaginary parallel worlds and escapism to flee from our reality doesn't reflect a degree of superiority but rather a hindrance in accepting what we face daily.

We have also lost the habit of sniffing each other to understand one another. Besides unpleasant odors, none of us would ever dream of sniffing another person during a work meeting. Life could be lived differently if we didn't need to write, daydream, go to the movies, watch television, and so on, or read books.

Our potentially limitless brain is our limit. We cannot live without these things, especially if we have an urbanized life. I would find it very difficult myself to not do anything like any animal. To dedicate myself to eating, sleeping, defecating, and having sex. But when I think about it, personally, these four actions are precisely what give us a sense of fulfillment and true happiness. Yet, in our modern world, we sleep poorly and little, eat junk food, and sex is experienced in extreme ways (either too strange or too little or is a social taboo). The only animalistic action left to us is sitting on the toilet. No one can take this away from us; everything ends in poop: a meager consolation.

5 AN ANIMAL DOCTOR

In the grand theater of life, where humans often take center stage, it seems that our non-human companions have honed their diagnostic skills behind the scenes. Yes, you heard right: your four-legged friend might be the next Florence Nightingale, with a nose that rivals any high-tech medical device. The world has had its share of famous animals with extraordinary disease-detecting abilities, and it appears that Churchill's cat had more to offer than just warming the Prime Minister's lap. Let's embark on a journey into the fascinating world of animal diagnostics, where our beloved pets play the role of undercover health detectives.

It all began with Sir Winston Churchill (The International

Churchill Society, n.d.)[3], the man whose political savvy was matched only by the wisdom of his feline companion. Rumor has it that Churchill's cat, a humble feline with a knack for political purring, had an extraordinary talent for sniffing out illness. Some may dismiss this as mere coincidence, attributing the cat's behaviour to a love of warm laps and cozy blankets. But what if we dared to entertain the idea that cats, with their aloof and nonchalant demeanor, are secretly studying human physiology? Perhaps they conduct their own version of medical research during the stealthy hours of the night. The next time your cat curls up beside you, consider it a health check-up: a gentle reminder to book that long-overdue doctor's appointment.

Meanwhile, dogs refuse to be outdone in the realm of medical marvels. Forget expensive diagnostic tests; your loyal four-legged friend might be the health superhero you never knew you needed. Research has shown that dogs have an impressive ability to detect certain types of cancer with their superior sense of smell (Willis et al., 2004)[4]. Picture this: you're lying on the couch, binge-watching your favorite TV series, when Fido starts acting strange. Before dismissing it as an attention-seeking stunt, consider the possibility that he's trying to communicate a vital message. It turns

---

[3] The International Churchill Society. (n.d.). *Pets & Animals*. Retrieved November 28, 2025, from https://winstonchurchill.org/the-life-of-churchill/pets-animals/.

[4] Willis, C. M., Church, S. M., Guest, C. M., Cook, W. A., McCarthy, N., Bransbury, A. J., Church, M. R. T., & Church, J. C. T. (2004). Olfactory detection of human bladder cancer by dogs: a proof-of-principle study. The BMJ, 329(7468), 712.

out dogs have a knack for detecting cancer, and their alert behaviour might be a subtle signal to visit your doctor.

In a world dominated by technology and clinical precision, it's almost quaint to think that our animal friends could be the unsung heroes of healthcare. Imagine a scenario where our four-legged companions are stationed at every airport, train station, and shopping center, diligently sniffing out infectious diseases as a kind of furry front line. Forget intrusive swab tests and thermometers: our trusty animal allies could offer a more organic, albeit unconventional, solution. Picture a border security checkpoint with a lineup of cats and dogs, each wearing a stethoscope around their necks, ready to sniff out health risks. Who needs costly medical devices when you have a parade of pets ready to put their diagnostic talents to the test?

Let's not overlook our feathered friends, who, despite lacking the complicated mammalian sense of smell, can contribute to the world of healthcare. Take the case of the canary in the coal mine, a historical example of avian insight that saved countless human lives. Canaries, with their delicate constitutions, were used in coal mines to warn miners of toxic gases. If the canary succumbed to the fumes, miners knew it was time to evacuate. While it may seem like a grim practice of the past, it underscores the idea that animals, with their heightened sensitivity, can serve as our early warning system for health hazards.

As we dive deeper into the world of animal diagnostics, it's essential to acknowledge the extraordinary potential of our pets' sharp senses. Take viruses, for instance. While scientists work tirelessly in labs to develop tests and vaccines, our pets may be quietly waging a war against viruses on the home front. Consider the stereotypical image of a cat disdainfully turning away from a plate of food. Most have witnessed this scene, interpreting it as a feline display of picky eating. But what if your cat is actually a vigilant guardian of your well-being? Perhaps that rejected meal harbored microscopic invaders, and your cat, in a heroic act of dietary discernment, saved you from potential food poisoning.

Dogs, too, may be fighting a silent battle against viruses. Have you ever noticed your pup becoming unusually affectionate when you're unwell? Rather than chalking it up to simple sympathy, consider the possibility that your dog is an undercover virologist, detecting subtle changes in your scent that signal the onset of illness. It's like having a living, breathing virus detector by your side, ready to dispense cuddles and tail wags as a natural remedy.

Now, let's turn to the enigmatic world of psychic parrots. Though these feathered beings are often associated with mimicking human speech, there's more to them than meets (or mimics) the eye. Imagine a parrot perched on your shoulder, casually echoing your coughs and sneezes. Before dismissing it as mere mimicry, consider the alternative: your parrot may be a clairvoyant medical marvel, accurately predicting the onset of respiratory ailments. Forget the traditional thermometer; a psychic parrot might be the cutting-edge health tool you never knew you needed. As you sip herbal tea, lamenting the arrival of flu season, your feathered friend could offer preventive squawks, warning you to boost your immune system. Who needs a crystal ball when you have a parrot with an impeccable track record of forecasting your health woes?

While some may scoff at the notion of attributing medical insight to animals, there's a deep element of trust that exists between humans and their pets. This bond, built over centuries of coexistence and companionship, suggests that our domestic friends may indeed possess an innate understanding of our well-being. Take the story of Ruby (ABC15 Arizona, 2018)[5], the golden retriever who went viral for her alleged ability to detect changes in blood sugar levels. Ruby's owner, a diabetic, claimed that the dog could foresee hypoglycemic episodes before they occurred. Skeptics may argue that this is mere anecdote, but perhaps it's time to give credit where credit is due. Who needs continuous glucose monitoring when you have a loyal canine companion ready to bark the alarm at the first hint of a metabolic imbalance?

Expanding the menagerie of animals contributing to the healthcare field, let's dive into the world of marine mammals. Cetaceans, known for their intelligence and echolocation abilities, have also been studied for their medical diagnostic potential. Researchers have explored the idea that these sea's intellectuals may detect certain diseases through their echolocation clicks, which bounce off objects and return as echoes, giving them detailed

---

[5] ABC15 Arizona. (2018, November 14). *East Valley family says their dog can detect dangerous blood sugar levels*. YouTube. Retrieved November 28, 2025, from https://www.youtube.com/watch?v=W+n7-5_3psA

information about their surroundings. While this area of research is still in its infancy, the prospect of marine mammals aiding in medical detection is both fascinating and promising.

Moving from the ocean depths to the skies, birds of prey offer another intriguing example of animals with remarkable diagnostic abilities. Hawks, renowned for their exceptional vision and hunting prowess, have been trained to detect specific pathogens in agricultural settings. With their keen eyesight, hawks can spot signs of disease in other animals or crops, alerting farmers to potential outbreaks before they escalate. This proactive approach not only helps protect agricultural resources but also highlights the potential of avian species in playing a role in disease surveillance and prevention.

On land, beyond the realm of traditional pets, creatures like rats have demonstrated their disease-detecting capabilities. African giant pouched rats, for instance, have been trained to detect tuberculosis (TB) in human sputum samples with impressive accuracy (Poling et al., 2010)[6]. These rodents undergo rigorous training to identify the scent of TB bacteria, allowing them to analyze samples far more quickly than conventional methods. Their ability to sniff out TB could prove invaluable in regions where access to medical facilities and diagnostic equipment is limited,

---

[6] Poling, A., Weetjens, B., Cox, C., Beyene, N., & Sully, A. (2010). Using Giant African Pouched Rats (*Cricetomys gambianus*) to Detect Tuberculosis in Human Sputum Samples: 2009 Findings. *The Pan African Medical Journal*, 7, 17.

offering a cost-effective and efficient alternative for early detection and treatment.

In the vast and diverse tapestry of the animal kingdom, from cats and dogs to odontocetes and rats, there exists a wealth of untapped potential for advancing medical diagnostics. While human ingenuity has led to incredible advancements in technology and scientific research, it's important not to overlook the innate abilities of our animal counterparts. By recognizing and harnessing the unique skills of various species, we can pave the way for innovative healthcare approaches that complement traditional methods and improve outcomes for both humans and animals. So, the next time your pet exhibits unusual behaviour, consider it a subtle reminder of their hidden talents as health detectives in the grand theater of life.

To conclude, as we navigate the maze of medical marvels, let's not underestimate the potential of our animal companions. From cancer-sniffing dogs to psychic parrots and food-poisoning-preventing cats, our pets may be the unsung heroes of the healthcare system. While the scientific community continues its quest for groundbreaking discoveries, let's not forget the healing power of a wet nose, a warm purr, or a feathered premonition. So, here's to our non-human colleagues: the diagnosticians with tails and beaks, the healers with fur and feathers. In a world that sometimes overwhelms clinics, let's embrace the idea that our pets, with their quiet charm and silent wisdom, might just be the remedy we didn't know we needed—a prescription for joy, companionship, and, who knows, perhaps even a touch of psychic healthcare.

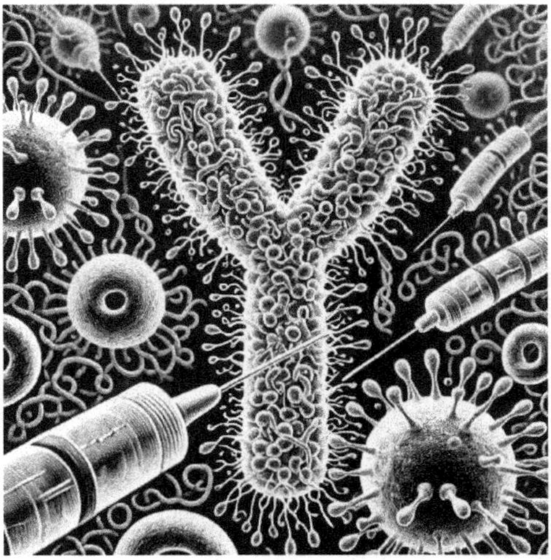

6 AND WHAT ABOUT ANTIBODIES?

In the complex dance of existence, antibodies emerge as the silent heroes of nature, a stark contrast to the often perplexing choices made by their human counterparts. While Homo Stultus immerses itself in activities as baffling as reality shows and debates about pineapple pizza, antibodies diligently play their role as vigilant guardians of nature, ready to protect their host from microscopic invaders.

The human immune system, a labyrinthine network of defenses, is often engaged in a ceaseless battle against its own kind—viruses, bacteria, and other infiltrators. In this microscopic battlefield, antibodies rise as indomitable defenders, endowed with an unparalleled ability to distinguish between allies and enemies, a capacity that escapes certain members of the human species in their often bewildering daily interactions.

Imagine, for a moment, a world where individuals possessed the sharp discernment of antibodies. Picture a realm where humans effortlessly identify harmful influences and toxic relationships, just as antibodies distinguish between pathogens and the body's cells. Alas, reality paints a different picture, where humans often embrace toxicity with open arms and share their living spaces with metaphorical pathogens.

The humble antibodies refrain from engaging in self-destructive behaviours or succumbing to toxic relationships. They do not give in to peer pressure, do not indulge in harmful habits, nor waste time on futile efforts. No, antibodies remain focused on what truly matters—safeguarding the well-being of the body. If only humans could channel a fraction of that clarity, the world might be a more harmonious place.

Contemplate the remarkable adaptability of antibodies, continually evolving to outsmart the cunning tactics of invading microorganisms. In the animal kingdom, survival of the fittest is not just a motto but a way of life. Animals adapt, evolve, and overcome challenges—a sharp contrast to certain human behaviours that can only be described as, well, less than intelligent.

Antibodies exemplify a level of cooperation and unity from which humans could learn. These microscopic soldiers form alliances, communicate seamlessly, and collaborate without the need for exhausting negotiations, lawyers, or heated debates. In the animal kingdom, survival is based on teamwork and cohesion, a valuable lesson for a species that often struggles to cooperate and unite on a large scale.

Antibodies, with their impeccable memory, learn from past experiences to better defend against future threats. If only humans demonstrated a similar ability to learn from history, the world would not be plagued by recurring conflicts and avoidable crises. The capacity to adapt and evolve mentally, emotionally, and socially is a trait antibodies possess in abundance—a trait that could undoubtedly benefit humankind.

Let us delve further into their complexity, peeling back the layers of their microscopic existence to uncover the wisdom inherent in their actions. Beyond their specificity, antibodies display an extraordinary diversity that contributes to their effectiveness. The human body produces a vast range of antibodies, each designed to recognize different antigens. This diversity ensures that collectively, antibodies can address a wide array of potential threats. It's a lesson in embracing differences and recognizing the value of diversity—a principle that, if applied socially, could lead to a richer and more resilient human experience.

Consider the adaptability of antibodies in the face of evolving threats. These microscopic warriors do not rest on their laurels; they evolve, adjusting their strategies to overcome the ever-changing tactics of invading microorganisms. The animal kingdom thrives on adaptability—a sharp contrast to certain human tendencies to resist change and cling to outdated ideologies.

In their cooperative efforts, antibodies act as role models for a human society that often grapples with division and discord. What if individuals could learn from the microscopic cooperation within their bodies and apply that spirit on a global scale? The lessons of unity and collaboration embedded in the actions of antibodies could potentially pave the way for a world where shared goals prevail over small differences.

Humans, as advanced as they are, often exhibit reactive toughness rather than proactive indomitability. They respond to crises with extraordinary tenacity, but true potential lies in the ability to anticipate challenges and adapt in advance, just like antibodies. If human society could embrace a more proactive buoyancy, driven by adaptability and continuous learning, the future would be brighter and less vulnerable to recurring crises.

Humans can learn much from this natural symphony. Imagine a society where every individual recognizes their unique role and contributes to the common good, just as antibodies do. Collective hardiness emerges from the sum of the parts, where every individual action strengthens the stalwartness of the entire system. In this way, society could develop a robustness that transcends mere survival and elevates toward prosperity.

In our ongoing exploration, we cannot overlook the concept of balance, a fundamental feature of antibodies and the animal kingdom in general. Antibodies not only defend the body but also maintain a delicate balance between attacking pathogens and protecting healthy cells. This balance is crucial to avoid collateral damage and preserve the body's homeostasis.

Humans, on the other hand, often struggle to maintain a similar balance in their lives and societies. The pressures of modernity, conflicting interests, and the inability to harmonize development and conservation often lead to imbalances with devastating consequences. Learning from antibodies and nature could guide humanity toward a healthier and more sustainable balance, where economic development, social well-being, and environmental health support each other.

Antibodies also teach us the importance of altruism. In their relentless work to protect the body, they always place collective well-being above their own survival. This altruism is deeply ingrained in nature, where many species exhibit behaviours that promote the survival of their community or family.

Humans, with their social and cultural complexity, could benefit greatly from embracing a more altruistic ethic. In a world where selfishness and unchecked competition often dominate, examples of altruism in nature and antibodies can serve as a beacon. If human societies could prioritize collective well-being and altruistic cooperation, the possibilities for building a more just and sustainable future would be immense.

Let us also reflect on the concept of memory, an extraordinary feature of antibodies that recalls past experiences to better face future challenges. This principle of immune memory is a powerful parallel to human collective memory. The lessons learned from past experiences, whether successes or failures, are essential to avoid repetitive mistakes and to build a solid foundation for future progress.

However, human history is often marked by a tendency to repeat the same mistakes, ignoring the lessons of the past. Antibodies remind us of the importance of keeping the memory of experiences alive and using that knowledge to guide future actions.

7 HOW GOOD IS ARSENIC?

In the grand tapestry of life, humans have long considered themselves masters of the domain of toxic substances. We take pride in our ability to label substances as lethal, toxic, or dangerous, as though the rest of the animal kingdom is merely a spectator in the theater of chemical warfare. Oh, how naive we are to assume that what is deadly for us is equally dangerous for our furry, feathered, and scaly companions. Let's embark on a journey into the singular realm of toxic substances and discover the surprising differences between humans and animals.

Take arsenic, for example. Humans have dubbed it a sinister poison, a substance not to be trifled with. But in the parallel universe of animal physiology, arsenic seems to have lost its malevolent touch. Mice, those quiet little creatures that scurry about, appear to possess an incredible ability to resist the toxicity of arsenic as if it were just a light morning fog. Imagine the arrogance of our species, believing that arsenic, a well-known killer in our realm, is nothing more than a nuisance for the mice.

Perhaps we should envy these resilient rodents, who ingest arsenic without a care in the world. Are we, the supposedly superior beings, missing some secret elixir of life that the mice are sipping behind our backs?

Humans have painstakingly compiled lists of substances deemed harmful, with warnings adorning every corner. Yet, animals seem to enjoy a world full of what we call toxic cocktails. The lethal dose for humans becomes just an appetizer for some of our animal counterparts. Aside from the sarcasm, it's truly astonishing how animals navigate their environment so effortlessly when faced with substances that would send us rushing to the emergency room. A bit of cyanide here, a pinch of poison there—just another day in nature for creatures that seem impervious to the dangers we obsessively catalog. Speaking of poison, let's talk about ants. Those small, industrious insects, the unsung heroes of the microcosmic world, have venom that could be considered nectar of the gods by their standards.

While a single ant sting might be just an inconvenience for us, it's a strategic masterpiece for the ant kingdom. And yet, humans have managed to turn this venom into a commodity, extracting it for research purposes. Oh, the irony! We marvel at the resilience of these tiny architects, while our own skin rebels against a substance so small it requires a microscope to truly appreciate it. Perhaps the ants, with their infinite wisdom, have developed a secret venomous society to laugh at our vain attempts to understand their microscopic kingdom.

Let's venture into the realm of radiation, a force that sends chills down the spines of humans. Chernobyl and Fukushima remain grim reminders of our vulnerability to this invisible danger. Meanwhile, some animals, blissfully unaware of our nuclear concerns, thrive in these radioactive hot zones. The notorious mutant powers we associate with comic book characters seem to be the daily reality for certain resilient creatures.

Imagine this: a family of mutant mice living amidst the ruins of a nuclear disaster, seemingly indifferent to the radiation that would have our Geiger counters crying out in protest. Are they the superheroes we didn't know we needed, or are we merely the fragile mortals in a world where radiation is just another flavor in the cosmic buffet?

Let's not forget the darling of cosmetic procedures, the miraculous worker in the search for eternal youth: botulinum toxin. Humans inject it into their faces to defy wrinkles and expression lines, yet this potent neurotoxin is merely an informal side dish for certain animals. Some creatures feast on organisms laden with botulinum, indifferent to the lethal effects that would leave us paralyzed. Here we are, desperately injecting small amounts of a substance that could be a death sentence if ingested in larger quantities, while wild animals consume it as part of their everyday fare. It's almost comical how hard we strive for beauty, while other species flaunt their natural grace, untouched by the poison we so willingly embrace.

While we boast of our intelligence and technological prowess, there is a certain brilliance in the way animals navigate the toxic minefield that we've labeled the natural world. Perhaps, just perhaps, we could learn a thing or two from our non-human counterparts. While we fret about substances that could spell our doom, animals continue to enjoy a world where the term "toxic" is simply a matter of perspective. The next time we proudly declare a substance to be dangerous, let us remember that in the grand scheme of things, the target may be on us. It's not just about surviving in a toxic world; it's about thriving. Animals, unlike humans, have adapted over eons to coexist with substances that would send our scientific minds into a frenzy of caution.

The adaptability of these creatures is a testament to the wisdom inherent in the very fabric of their being. Consider the humble bacteria, those microscopic entities often at the center of our battles against infectious diseases. Some bacteria, inhabiting the harshest environments, have developed an almost incidental relationship with substances that are lethal to us. They thrive in extreme conditions, developing mechanisms to resist doses of toxins that would obliterate human cells. The arrogance of our species, with our antibacterial soaps and disinfectants, stands in stark contrast to the tenacity of these microorganisms. We wage war on bacteria, hoping to sterilize our surroundings, while they, in turn, laugh at our vain attempts to control a world they have silently dominated for eons.

Or take the complicated case of lead, a heavy metal we have diligently removed from gasoline and paints due to its well-documented toxicity in humans. Yet, birds seem to have missed the memo.

The sky's navigators, ingesting lead ammunition left by hunters, show remarkable resilience to levels that would turn us into cautionary tales. Birds of prey, in particular, have become the unwitting subjects of our chemical experiments. The lead ammunition fired into the air, aimed solely at prey with no thought for ecological consequences, ends up in the digestive systems of these majestic hunters (American Eagle Foundation, n.d.)[7]. Yet, they fly above us, seemingly immune to the dangers we have unleashed upon them. The irony here is palpable. We ban lead from our environment to protect ourselves, while the sky's guardians continue to ingest it with a carelessness that mocks our futile attempts at environmental management.

Diving into the aquatic realm, a new dimension of chemical fortitude is revealed. Fish, the elegant inhabitants of the depths, swim through waters we have polluted with industrial pollutants and agricultural runoff. Substances like mercury, once a stern warning for pregnant women in our world, has become an integral part of the

---

[7] American Eagle Foundation. (n.d.). *Lead Toxicity in Bald Eagles*. Retrieved November 28, 2025, from https://eagles.org/lead-toxicity-in-bald-eagles/

fish feast beneath the surface.

While we meticulously monitor mercury levels in our seafood, the fish themselves seem to have evolved intricate mechanisms to minimize the toxic effects, however it does not mean they are completely immune from it. Our cautionary tales about mercury toxicity become fish tales, the aquatic equivalent of spooky stories around a campfire meant to scare small fish into good behaviour.

And let's not forget the flora in this toxic tale. Plants, often silent observers in the drama of life, have their arsenals of chemicals that would unsettle our pharmaceutical labs. From alkaloids to tannins, plants have concocted a vast range of compounds to deter potential herbivores. Yet, animals have not only adapted to these chemical defenses but have even developed a taste for the very substances we consider toxic. Go figure! The same plant poisons that deter herbivores become seasoning for some animal species. It's as if nature is playing a cosmic joke, challenging our perception of what is harmful and what is beneficial.

As we marvel at the resilience of animals in the face of natural toxins, the ultimate irony lies in the toxic legacy we, as humans, have left in our wake. Synthetic chemicals, born in the furnace of our industrial might, now permeate the environment in ways our ancestors could not have imagined.

Plastic, a marvel of modern convenience, infests the oceans and infiltrates the food chain. Pesticides, designed to protect our crops, end up in the bodies of unwitting creatures. Hormone-disrupting chemicals alter wildlife populations. In this brave new world of human-made toxins, animals face challenges far beyond the natural order. The durability that once equipped them to deal with ancient toxins is now being tested against a flood of chemicals that nature never anticipated. From this emerges a humbling lesson. Our perception of toxicity is a construct of our own vulnerabilities. What is lethal to us is but a puzzle for the other inhabitants of this planet. Perhaps it is time to abandon our arrogance and recognize that the world does not revolve around the human experience.

The tenacity of animals in the face of what we consider toxic challenges our understanding of the natural order. It is a reminder that life, in all its diverse forms, has found its balance in the intricate dance of survival. So, the next time we label a substance as toxic, let us do so with a touch of humility. After all, in the grand theater of life, the applause doesn't belong only to the leading actors; sometimes, it's the extras, the supporting cast of resilient creatures, who steal the show.

In our pursuit of progress and domination over the natural world, we often overlook the invisible cost of our arrogance. The very toxins we've unleashed on the planet, confident in our ability to control and manipulate, now return to haunt us in unexpected ways. From the microplastics that permeate the depths of the ocean to the invisible specter of air pollution blanketing our cities, we have created a toxic legacy that affects not only us but also the countless creatures with whom we share this Earth.

Consider the tribulation of urban animals, those resilient

beings trying to navigate the concrete jungles filled with pollutants. Squirrels darting between skyscrapers, birds building nests on ledges, and insects finding refuge in the tiniest patches of green – all are confronting a toxic landscape we've built in the name of progress. The irony is palpable: the very beings we consider pests and nuisances are the unsung heroes, adapting to a world we've shaped without considering the intricate web of life that sustains us all. The toxic fumes we emit from our factories, the exhaust from our vehicles, and the chemicals we carelessly release into the air are the invisible foes these urban warriors face. Amid our ignorance, some creatures suffer more than others from the consequences of our toxic ventures.

Amphibians, delicate beings that inhabit both land and water, are facing a crisis that reflects the fragility of our interconnected ecosystems. Pesticides, pollutants, and habitat destruction converge to create a deadly brew for these animals, leading to an unprecedented decline in amphibian populations worldwide. The silent suffering of frogs, salamanders, and newts is a warning sign we cannot afford to ignore. The disappearance of amphibians foreshadows a broader collapse in biodiversity, a cascading effect that could eventually reach our doorstep. Yet, in our anthropocentric arrogance, we continue to ignore the fate of these amphibious sentinels as inconsequential.

The intricate dance between predators and prey takes on a new dimension in nature. Predators, equipped with a series of adaptations that allow them to capture prey, become unwitting victims of the very substances that saturate their hunting territories. The situation of charismatic carnivores like eagles, bears, and big cats, facing the toxic consequences of consuming contaminated prey, is a harsh reminder of the delicate ecological balance we walk.

Consider the bald eagle, a symbol of freedom and majesty in the skies. Despite significant conservation efforts, these regal birds find themselves struggling with a dark adversary: persistent organic pollutants. The fish they catch from rivers, full of chemicals we've introduced thoughtlessly, becomes a Trojan horse of toxic fate.

As we scrutinize the intricate dynamics of toxicity in the natural world, we call for reflection on our place within this vast ecosystem. Our actions, often driven by short-term gains and blind ambitions, reverberate through what we call life. The toxins we release, both intentionally and unintentionally, become the silent orchestrators of a symphony in which we are mere players. In our pursuit of dominance, let us not forget that we share this planet with resilient beings who face our toxic legacy. Perhaps it's time to shift our perspective, to recognize that the consequences of our actions extend far beyond our immediate concerns. The toxic truths of our existence are not confined to the animal kingdom; they infiltrate the very fabric of our interconnected world, demanding a collective awakening to the consequences of our shared journey on this pale blue dot we call Home.

8 LET'S SAVE OURSELVES: WHEN NATURE GETS A HEAD START ON HUMANITY

Let's take a break from the seriousness of our Non-Human companions for a chapter and introduce a modest series of examples on how we, Stupid Humans, have missed the point.

The Chameleon's Camouflage: Chameleons blend into their environment to escape predators and surprise prey, a technique that humans have perfected in trying to avoid uncomfortable commitments or by popping up as "social chameleons" in various social groups to be accepted everywhere. Instead of iridophores, we use excuses and white lies like "I need to check my schedule" or "I lost connection." The chameleon's ability to change color and skin structure to perfectly match its environment is something humans try to emulate with clothing, makeup, and even social media filters, hoping to go unnoticed or impress others.

The Science Behind the Chameleon's Camouflage:

Chameleons have special cells called chromatophores and iridophores, which contain pigments and reflective structures, allowing them to change color. This phenomenon is controlled by nervous and hormonal signals that regulate the expansion and contraction of these cells to achieve the desired color.

The Regenerative Powers of the Axolotl: The axolotl can regenerate limbs and tissues, an ability that would come in handy for humans, especially after a night of excess that leaves not only physical but also moral hangovers. Instead of regenerating like these amphibians, we rely on post-bender recovery treatments and technologies like Photoshop to fix image damage on social media. Scientists are trying to decode the axolotl's genetic code to apply it to human regenerative medicine, but until then, we settle for band-aids and long waits for healing.

The Regeneration Process in Axolotls: Axolotls can regenerate limbs, hearts, spines, and parts of their brains thanks to the presence of pluripotent stem cells. These cells can differentiate into various types of tissue, allowing for complete and functional regeneration. Researchers are studying the molecular and genetic mechanisms that enable this regeneration, with the aim of applying them to human medicine.

The Language of Bees: Bees dance to communicate complex information about food sources, a skill that humans awkwardly try to imitate during work meetings or in social clubs, but with less grace and precision. Imagine if our "social dances" were as efficient and precise; we'd save a mountain of unnecessary emails and messages. Bees regulate the duration and intensity of their dances based on environmental factors, while humans often misunderstand simple signals like "I'm fine," which could mean "help me" or "leave me alone."

The Bee Dance: Worker bees communicate with each other through a series of movements called the "waggle dance" or "abdomen dance." This behaviour allows bees to indicate the direction and distance of food sources relative to the hive. The dance includes a figure-eight path with vibrating abdomen movements that transmit precise information about the location of resources.

Dolphin Echolocation: delphinids use echolocation to navigate and communicate, while we humans, armed with cutting-edge technology, still end up losing our car keys or having misunderstandings on a famous instant messaging service. If only we were as skilled at using our resources, perhaps we'd avoid 99% of daily accidents. These Pelagic sentinels adjust the frequency and direction of their clicks to adapt to their environment, showing a flexibility that we humans can only dream of, especially when following GPS directions that lead us into a dead-end street.

The Mechanism of Echolocation in some marine mammals: Cetaceans emit a series of high-frequency clicks, which bounce off objects and return as echoes. By analyzing the return time and the quality of the echoes, these social swimmers can create a detailed image of their environment. This system is so precise that these aquatic acrobats can identify the shape, size, and even the composition of materials of surrounding objects.

The Intelligent Crow: Crows use tools and pass on knowledge across generations, a skill often lacking in humans, who repeat the same historical mistakes and struggle with the challenge of "passing the torch." While crows learn and innovate, humans often forget lessons already learned and rely on instruction manuals for every new challenge. Crows demonstrate problem-solving abilities that challenge our intelligence, especially when we get lost in the maze of instructions for assembling a famous brand piece of furniture.

The Use of Tools in Crows: New Caledonian crows are known for their ability to create and use tools to obtain food. They can shape twigs and leaves to build hooks and probes, demonstrating a high level of cognition and problem-solving. Moreover, crows pass these skills down to their young, showing a complex and developed animal culture.

The Engineering of Spider Webs: Spiders weave webs with engineering precision, creating strong and functional structures. Meanwhile, humans struggle with furniture and constructions that collapse after a decade. If only we could imitate their efficiency, we might have fewer problems with home maintenance and infrastructure. Spider webs, with their strength and flexibility, inspire innovations in materials, but humans still seem unable to create something equally effective without a team of engineers and an astronomical budget.

The Construction of Spider Webs: Spider webs are built using silk produced by the spiders' silk-producing glands. This silk is extraordinarily strong and flexible, able to absorb large amounts of energy without breaking. Spiders weave their webs following complex geometric patterns that maximize efficiency in capturing prey, demonstrating a mastery of natural engineering.

The Mastery of Hibernation in Bears: Bears meticulously prepare for hibernation by accumulating reserves and optimizing energy consumption. Humans, on the other hand, always find themselves rushing at the last minute, whether preparing for an exam or meeting a work deadline. Instead of accumulating fat, we accumulate stress and sleep debt. Bears enter a hibernation state that conserves energy efficiently, while we rely on coffee and power naps to survive increasingly hectic days. During hibernation, bears drastically reduce their metabolism, lower their body temperature, and slow their heart rate. This allows them to survive for months without eating, using the fat reserves accumulated during the summer and fall. Bears periodically wake up for brief periods during hibernation, but they generally remain in a state of deep torpor until spring.

The Endurance of Tardigrades: Tardigrades, a phylum of eight-legged segmented micro-animal, survive the most extreme conditions thanks to genetic adjustments, while humans, faced with a crisis, often retreat into binge-watching Netflix and comfort food. If only we had a fraction of their toughness, perhaps we would handle daily challenges better without relying on fleeting remedies. Tardigrades thrive in environments where humans couldn't survive for even a day, and their ability to enter a state of suspended animation makes us reflect on our own biological limits.

Tardigrades, or "water bears," can enter a state called cryptobiosis, where their metabolism nearly stops completely. In this state, tardigrades can survive extreme temperatures, high pressures, intense radiation, and a lack of water for years. When environmental conditions improve, tardigrades "come back to life," resuming their normal functions.

The Falcon Peregrine's Flight Feat: Peregrine falcons are masters of high-speed and precise flight, while humans are still struggling with road traffic and flight delays. If we could fly with their skill, we might avoid the daily frustrations of commuters and travelers. The peregrine falcon's ability to reach impressive speeds without losing control is a far-off dream for humans, who often fight against gravity with slow and inefficient means of transportation. The peregrine falcon can reach diving speeds of over 320 km/h(198.8442 MPH), making it the fastest animal in the world. During hunting, the falcon uses its speed and agility to catch prey mid-air. Its wing structure and respiratory system are optimized to sustain high speeds and rapid maneuvers, making it a formidable predator.

The Mysterious World of Protists: Protists play vital roles in global ecosystems, demonstrating surprising complexity. Humans, despite their claim to superiority, often overlook the importance of essential ecological balances, preferring to focus on problems they themselves have created. Perhaps we should learn from protists the art of symbiosis and sustainability. These microscopic organisms, though invisible to the naked eye, are essential for life on Earth, while we humans, in our arrogance, often forget the importance of the small cogs that make the great machine of nature work.

The Role of Protists in Ecosystems: Protists include a wide range of simple unicellular and multicellular organisms that play crucial roles as primary producers, decomposers, and parasites. They contribute to the nutrient cycle, produce oxygen, and form the base of the food chain in many aquatic ecosystems. Their diversity and adaptability make them fundamental to the stability and productivity of natural environments.

In these brief glimpses, we have only scratched the surface of the extraordinary diversity and complexity of life on Earth. Every organism, large or small, offers a unique opportunity to explore the mysteries of evolution, resilience, and nature's ingenuity. As scientists continue to unravel the biological wonders around us, we find ourselves immersed in a world of ongoing discoveries, where each new insight brings us a little closer to understanding the deepest secrets of life. However, we can't help but notice that despite all our technology and innovation, there are still fundamental lessons to be learned from our planetary companions.

9 EXISTENTIAL PROBLEMS

Once upon a time, in a world not so far away, there existed a species known as humans. These peculiar beings considered themselves the crème de la crème of the animal kingdom, boasting of their intelligence and self-awareness. Little did they know, the animal kingdom had its own interpretation of intellect—which didn't involve long philosophical debates or existential angst. Let's dive deeper into the whimsical menagerie that is the zoo of absurdity, exploring more examples that highlight the stark contrasts between human complexities and the simplicity of the animal realm.

DEEP THOUGHTS
vs
SIMPLE PLEASURES

Humans, with their oversized brains and tendency to ponder the meaning of life, often find themselves tangled in a web of existential angst. The intricate weave of human thought, interlaced with threads of self-doubt and metaphysical reflections, stands in sharp contrast to the simplicity of animal joy. Think of a cat basking in a sunbeam, blissful and unbothered by the Descartes' reflections or the existential dread explored by Sartre.[8] No existential angst lingers in those sparkling eyes—only pure joy.

Imagine a world where humans, instead of grappling with the complexities of consciousness, found comfort in the rhythmic joy of everyday life—a world where contemplating the intricacies of existence was replaced by a good game of fetch in the backyard. What a world that would be, where the pursuit of happiness was as simple as chasing a squeaky toy. Picture a rabbit nibbling on fresh grass, savoring the simplicity of the present moment, unburdened by the weight of existential reflection.

---

[8] Descartes, R. (1641). *Meditations on First Philosophy*. Michel Soly. Sartre, J. P. (1943). *Being and Nothingness: An Essay on Phenomenological Ontology*. Gallimard.

VERBAL DIARRHEA
vs
ACTION-ORIENTED TRANSMISSION

Humans have a unique talent for creating complex languages, engaging in endless chatter about the past, present, and future. The art of language has allowed humans to discuss matters ranging from the profound to the utterly trivial. They debate politics, dissect literature, and engage in conversations that span the full range of human experience. Meanwhile, animals communicate through actions and energy, relying on non-verbal language that doesn't require dictionaries or thesauruses. A pack of wolves, for instance, communicates through body language and vocalizations, seamlessly coordinating their actions without needing lengthy discussions about strategy. A hungry lion doesn't waste time discussing its hunting strategy; it simply pounces on its prey.

No need for elaborate speeches or digital presentations—just raw, decisive action. In the animal kingdom, there are no verbose Social Media Talks on the art of effective contact—just the practical application of passing time.

If only humans could adopt a similar approach, cutting through the linguistic noise and embracing the simplicity of action-oriented link! Imagine a world where the elaborate debates of Plato or Aristotle were replaced by the frankness of a bee's dance to communicate the position of nectar.

Picture a world where the acrobatic display of a dolphin conveyed more than a thousand words ever could.

EMOTIONAL BAGGAGE
VS
LIVING INSTINCTIVELY

Humans carry the weight of emotional baggage, from love disappointments to mid-life crises. The human psyche, a tangled mess of emotions and experiences, often leads to rollercoaster rides of feelings that range from euphoria to despair. Therapy sessions, self-help books, and endless introspection have become the tools by which humans attempt to navigate the labyrinth of their minds. Animals, on the other hand, live according to their instincts. They don't dwell on yesterday's mishaps or worry about tomorrow's uncertainties. A gazelle doesn't spend its days fretting over embarrassing moments—it's too busy running from predators. The emotional baggage that weighs down human existence is practically nonexistent in the animal kingdom. Instead, animals operate on a direct spectrum of instincts that guide their actions and reactions.

If only humans could shed the unnecessary layers of emotional complexity and embrace the simplicity of instinctual living. Consider a meerkat, standing guard with focused attention, unburdened by the psychological complexities explored by Freud or Jung. Imagine a world where humans, like elephants recalling ancient paths, navigated life with a sharp survival instinct.

Yet, our psychological complexity also offers a gift to life. It allows us to become aware of life—and to be alive—rather than just existing. That awareness holds meaning for life, not just for humans.

COMPLICATED RELATIONSHIPS
VS
SIMPLE BONDS

Humans navigate a labyrinth of complex relationships, full of emotions and social dynamics that would confuse even the most attentive observer. Human relationships often involve a delicate dance of social etiquette, emotional intelligence, and interpersonal skills. Animals, on the other hand, keep things simpler. If a cat likes you, it rubs against you and purrs. If not, you might need a few band-aids. No need for relationship gurus or dating apps—just a direct expression of affection or disdain. The simplicity of animal bonds, forged through instinct and genuine emotion, stands in stark contrast to complicated human relationships. Imagine a world where romance was as simple as the loyalty of a wolf to its pack or the constant love of a dog for its human companion.

No need for relationship advice or online dating algorithms—just the authenticity of simple connections that transcend the need for complex social norms. Imagine a world where a pair of swans, in their elegance, communicated the beauty of commitment without the need for complicated relationship status updates.

ARTIFICIAL PROBLEMS
vs
REAL-LIFE CHALLENGES

Humans excel at creating artificial problems—complex puzzles and challenges that often lead to self-inflicted stress. The human mind, a labyrinth of thoughts and desires, has a talent for conjuring scenarios that provoke anxiety and worry. Animals, on the other hand, face real-life challenges such as finding food, avoiding predators, and marking territories. The simplicity of animal existence lies in their ability to confront tangible and immediate challenges without the burden of abstract, self-created problems. In a world where humans could focus on real-life challenges with the same urgency and efficiency as their animal counterparts, perhaps the unnecessary stress and anxiety that permeate human existence would dissipate. Instead of struggling with artificial dilemmas, humans could channel their energy into tackling the tangible obstacles in their path.

Imagine a world where the thoughts of Kierkegaard or Camus were replaced by the practical wisdom of a beaver building a dam to protect its habitat. Consider a world where a spider's expertise in weaving webs symbolized problem-solving in its purest form.

TECHNOLOGICAL OVERLOAD
VS
HARMONY OF NATURE

In the human realm, technology reigns supreme. The constant hum of electronic devices, the glow of screens, and the obsessive algorithms have become integral parts of daily life. Humans take pride in their technological advancements, from smartphones that unlock with a mere glance to artificial intelligence that can simulate a conversation. Compare this to the serene existence of animals immersed in the harmony of nature. Birds build nests with twigs, and ants create intricate colonies—all without needing a single line of code. Imagine a world where humans exchanged their gadgets for the simplicity of a bird's song or the rustling of leaves in the wind. The technology of nature, free from glitches and updates, offers a timeless connection that transcends the digital complexities of human existence.

OBSESSION WITH TIME
vs
THE ETERNAL PRESENT

Humans are often slaves to time, meticulously planning their days, weeks, and years. Clocks tick in the background, deadlines loom, and the concepts of past, present, and future dominate human consciousness. The constant pursuit of productivity and the fear of wasted time create an endless cycle of stress and restlessness. Contrast this with animals, who live blissfully in the eternal present. A squirrel collects nuts without worrying about tomorrow, a bear hibernates in winter without agonizing over lost time, and a butterfly dances from flower to flower, savoring every moment. Imagine a world where humans embraced the simplicity of living in the present, free from the shackles of time-induced anxiety. Perhaps, in such a world, the relentless pursuit of productivity would give way to a genuine appreciation for the beauty unfolding in every passing moment.

MATERIAL OBSESSION
VS
MINIMALIST LIVING

Humans are sadly known for their materialistic pursuits, accumulating possessions in search of status, comfort, and validation. The relentless quest for more—more wealth, more goods, more experiences—often leads to a complicated and confusing existence. Compare this with the minimalist life of animals. A hermit crab carries its home on its back, a weaver spider spins a delicate web for shelter, and a beaver builds a dam with only the essentials. Imagine a world where humans found satisfaction in the simplicity of their environment, free from the constant desire for material accumulation. Instead of measuring success by the size of their homes or the brand of their cars, humans could find fulfillment in the elegant simplicity of a well-built nest or a cozy den.

ENVIRONMENTAL IGNORANCE
vs
NATURAL WISDOM

In their pursuit of progress, humans often overlook the impact of their actions on the environment. Deforestation, pollution, and climate change are just some of the consequences of human activities driven by short-term gains. Contrast this with the innate environmental wisdom of animals. They instinctively understand their ecosystems, adapting to seasonal changes and maintaining a delicate balance with nature. Imagine a world where humans surrendered to the natural wisdom of animals, living in harmony with the environment rather than exploiting it. Perhaps, by observing the sustainable practices of animals, humans could learn to tread lightly on Earth, ensuring a healthier planet for future generations.

In reflection, the zoo of absurdity invites us to reconsider the complexities of human existence in the context of the simplicity of the animal kingdom. So, the next time the weight of the world feels overwhelming, take inspiration from the simplicity of nature—embrace the moment, simplify your goals, and live in harmony with the world around you. After all, in the grand game of life, simplicity might be the thread that weaves the most beautiful stories.

10 A STORY WITHOUT GHOSTS

As the saga of human superiority continues, it becomes increasingly clear that animals possess a remarkable ability to selectively ignore. While humans boast about their ability to perceive the supernatural, animals are content to remain blissfully unaware, navigating the world with a pragmatic approach that mocks ethereal drama.

Take, for example, the small goldfish. In its aquatic home, it swims in circles, completely uninterested in the supposed ghostly apparitions that humans claim to see. Perhaps the goldfish has a deeper understanding of existence: why worry about the supernatural when there are aquatic plants to nibble on and bubbles to chase?

Through the vast African savannas, the mighty elephants

sound their disdain for the concept of spirits. With their enormous size and majestic tusks, elephants carry the weight of the world on their shoulders, utterly indifferent to the invisible specters with whom humans claim to communicate. After all, when you're a pachyderm, the trivial matters of the spiritual realm pale in comparison to the grandeur of a dust bath.

In the lush rainforests, monkeys nimbly leap from branch to branch, their playful antics suggesting a complete lack of interest in the spectral parodies of humans. While humans toil in search of the unseen, the monkeys enjoy the tangible delights of fruit, insects, and the occasional game of hide-and-seek among the leaves.

The majestic lion, king of the jungle, yawns indifferently at human concerns about ghosts and spirits. From the top of the food chain, the lion's perspective on the supernatural is one of regal apathy. After all, when you have a pride to lead and a savanna to rule, who has time for the ethereal whims of invisible entities?

Meanwhile, in the depths of the ocean, where sunlight struggles to penetrate the mysterious darkness, creatures like the abyssal fish emit a ghostly glow. Unconcerned with the human fascination with spirits, these denizens of the deep continue their activities in a realm where spectral apparitions seem a trivial concern compared to the mysteries hidden in the abyss.

Traveling to the Himalayan mountains, where the air is thin and the landscape inhospitable, snow leopards traverse the steep slopes with grace and stealth. These fascinating creatures, perfectly adapted to their harsh environment, seem untouched by the intangible fears that haunt human minds. The ghostly apparitions that humans claim to see may pale in significance compared to the snow leopard's survival in a place where every step could mean life or death.

In the heart of the American Midwest, where endless fields of corn stretch to the horizon, cows graze with quiet indifference to the spectral tales that captivate human communities. The cows, with their gentle behaviour and soulful eyes, seem to find comfort in the simplicity of the earthly realm, undisturbed by the ghost stories that permeate human folklore. Their serene existence raises questions about the nature of spirituality and whether the search for the invisible is a uniquely human endeavor.

The majestic humpback whales, with their haunting songs echoing through the deep oceans, seem to communicate with a mysterious underwater realm that eludes human understanding. While humans gaze at the night sky, seeking ethereal constellations and signs, the whales dive into the enigmatic abyss, their underwater symphonies resonating with a connection to a spiritual plane that transcends the human concept of ghosts. Perhaps, in the vastness of the ocean, the whales navigate a spiritual realm that humans can only glimpse in their dreams.

In the dense jungles of Southeast Asia, where the air is thick with humidity and the foliage teems with life, the mysterious Komodo dragon patrols its territory with a regal indifference to the invisible. With a prehistoric presence, the Komodo dragon embodies stoic resilience against the intangible fears that afflict human imagination. Their existence, marked by a primordial connection to the Earth, stands as a testament to the enduring power of the tangible over the ephemeral.

In the vast savannas of Africa, where the sun paints the landscape in warm tones, meerkats stand guard with an unwavering attention to the present moment. These small mammals, with their sharp observational skills, demonstrate a practical wisdom that surpasses the complexities of human spiritual contemplation. The collective vigilance of the meerkats against tangible threats suggests a focused approach to life, one free from the intangible fears that plague humanity's search for the invisible.

While individuals in the human world continue to fervently share their spectral encounters, exchanging stories of shadowy figures and unsettling apparitions, the collective headshake of the animal kingdom persists. Their indifference to the spectral whims of humans raises a profound question: do animals possess a wisdom that humans, in their spectral pursuits, have overlooked? Perhaps, animals, with their instinctive focus on the tangible and the immediate, have uncovered a secret to a serene life—one not burdened by the intangible fears that haunt human imagination.

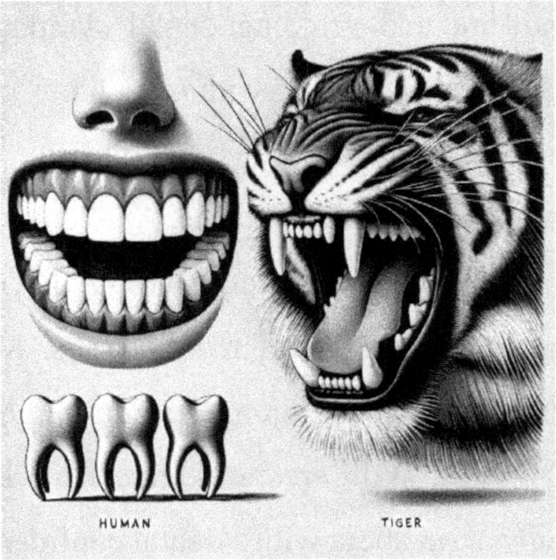

11 DENTAL DIVERSITY

In the grand theater of dental evolution, where every creature is an actor in the intricate dance of survival, humans often find themselves on the central stage, basking in the glory of their presumed superior teeth. But let us take a moment to abandon our anthropocentric ideas and embark on a sarcastic journey celebrating the wonderful dental diversity of the animal kingdom.

Ah, human teeth, meticulously polished, whitened, and flossed to perfection. We, Homo Stultus, proudly boast of our dental hygiene, using toothbrushes and floss as though our survival depends on it. But whom are we fooling? Our teeth are mere ornaments in the grand scheme of nature, toys that have been pampered into submission.

Our incisors, canines, molars—each assigned a specific task

in the great chewing and grinding dental symphony. We chew cooked meals, delicacies prepared with meticulous precision, while our canines, once weapons of carnivorous ferocity, are reduced to tools for tearing open stubborn packages. And let us not even begin to talk about our wisdom teeth—those vestigial reminders of our ancestors who had a bit more room in their skulls. Wisdom, indeed.

Now, let us turn our attention to the unsung heroes of dental evolution: the animals. Wild creatures, free from the shackles of toothbrush tyranny, strut about with a dental confidence that shames our human obsession with oral hygiene. Consider the beaver, for example. Armed with ever-growing incisors, these industrious architects can bite through trees with ease, creating dams and lodges that make our grandest buildings seem like child's play. No flossing required, thank you very much.

The shark, with its conveyor belt of teeth, takes a no-nonsense approach to dental maintenance. Lose a tooth? No problem! There's a real assembly line ready to take its place.

Humans, with our pitiful pair of two sets of teeth, might learn a thing or two about resilience from these ocean warriors. And what about the humble snail? A dental masterpiece, if such a thing can be said. The radula, a tongue-like organ adorned with thousands of tiny teeth, allows these slow and steady beings to feast on vegetation with a grace that escapes our masticatory abilities.

Let’s indulge in a little comparative analysis. Humans, with our delicate teeth, have the audacity to claim superiority in the chewing department. We cook our food, soften it to the point of mush, and then celebrate our exceptional chewing abilities. Meanwhile, the carnivores of the animal kingdom—lions and tigers, with their razor-sharp teeth—tear through raw meat with an elegance that makes our culinary skills seem laughable. Oh, the irony of our dental vanity! And let us not forget the self-proclaimed kings of dental perfection—orthodontists.

These brave individuals devote their lives to straightening human teeth, as though Mother Nature made a mistake in the grand design. Meanwhile, in the wild, animals parade proudly with a variety of dental configurations, a testament to the beauty of imperfection.

Let's rewind the tape and return to the days when our distant ancestors were still grappling with the mysteries of fire and stone tools. As the Homines Stulti made their entrance onto the stage of existence, our teeth were certainly not the polished jewels they are today. Oh no, they were sturdy, unpolished, and subject to the whims of natural selection. Imagine a time when our ancestors faced the world with a dental set that could compete with the most eccentric modern art installations.

Crooked incisors, protruding canines, and molars that seemed the result of a chaotic rock concert—these were the dental masterpieces that guided our journey through time. Now, compare this with the animal kingdom, where dental designs were finely tuned for the specific needs of each species. The saber-toothed tiger, for example, had fangs that could send shivers down even the bravest dentist's spine. These formidable canines weren't just for show; they were the keys to survival in a world where the line between predator and prey was as sharp as their teeth.

Then there were the herbivores, the gentle giants of the prehistoric savannas. Massive molars, designed to crush tough vegetation, allowed them to chew their way through the lush landscape without the need for culinary sophistication. No salads or smoothies for them—just raw, unadulterated foliage.

Fast forward to today, where humans have embraced the concept of oral hygiene with religious fervor. We brush, floss, and rinse with precision, as if our teeth were fragile artifacts in a slapstick museum. The animal kingdom, on the other hand, continues to thrive without the need for dental care regimens. Consider the hippopotamus, a creature renowned for its colossal tusks. These enormous incisors serve not only as formidable weapons but also as tools for stripping bark and branches. No dentist appointments, no fluoride treatments—just good old-fashioned tusks performing double duty in nature.

And what about the platypus, that strange creature from Australia? Armed with venomous spurs and a set of molars for grinding, this vicious monotreme navigates the aquatic world with a dental arsenal that puts our routines to shame.

And the Absurd Wisdom of Wisdom Teeth? Those latecomers to the tooth party who often make a dramatic entrance into our late adolescence or early twenties. Humans have considered these third molars as bothersome and useless, leading to a ceremonial ritual of extraction. Meanwhile, in the wild, animals like elephants not only boast a set of wisdom teeth but multiple sets throughout their life. These majestic giants, with their impressive dental line, never worry about crowded molars or impacted wisdom teeth. Nature, it seems, has a particular sense of humor.

We often boast about the power of our bite and the efficiency of our chewing process. But let's face it – compared to the animal kingdom, our bite is more of a nibble. Take the saltwater crocodile, for example, with a bite force that could rival the jaws of a hydraulic press. This prehistoric predator, armed with a dental arsenal designed for a life of aquatic domination, can crush the bones of its prey with a careless snap. In comparison, our feeble attempts to bite through a crunchy apple seem rather quaint.

And let's not forget the insects, the unsung heroes of the biting world. Ants, with their powerful mandibles, can carry objects many times their body weight. While we struggle with packaging, these little titans tackle obstacles with a determination that leaves our dental endeavors in the dust.

In the grand spectacle of existence, it's time for humans to embrace the diversity of dental designs that nature has bestowed upon its countless inhabitants. Our obsession with straight, white teeth might make for a charming Hollywood smile, but it's the dental oddities of the animal kingdom that truly steal the show. So, let's symbolically raise a glass to beavers, sharks, snails, saber-toothed tigers, hippos, platypuses, elephants, crocodiles, ants, and every other creature with a dental tale to tell. In the comedy of life, every toothy character plays a crucial role, reminding us that the stage is set not just for humans, but for every actor in the grand theater of nature's dental extravaganza.

Now, let's dive into the whimsical world of childhood fantasies and ponder the dilemma that plagues the legendary Tooth Fairy. As she flits from pillow to pillow, exchanging lost teeth for shiny coins, one can't help but wonder: does she prefer the meticulously brushed teeth of human children, or does she harbor a secret admiration for the sturdy teeth of the animal kingdom? Imagine her dilemma as she encounters a smile with empty spaces, evidence of a tooth lost in the murky depths of childhood mischief. Will she leave a token of appreciation for the child who diligently brushed and flossed, or will she be tempted to leave an extra reward for the little one who lost a tooth in a brave battle with an apple, like a small David facing a dental Goliath? Perhaps the Tooth Fairy, in her infinite wisdom, sees beyond the shine of human enamel and recognizes the intrinsic beauty in every tooth, whether it belongs to a child, a beaver, or a shark. After all, in the magical realm of childhood dreams, diversity is celebrated, and every tooth holds a story waiting to be told.

From the humble snail to the mighty crocodile, each tooth stands as a testament to the ingenuity of evolution and the boundless creativity of nature. So, the next time you find yourself admiring your smile in the mirror, take a moment to reflect on the wonderful variety of teeth that share this world with you.

And so, dear reader, as we draw the curtains on this whimsical tale of dental enchantment, let us carry with us the lessons learned from the toothy characters who graced the stage. In the end, it is not the perfection of our smiles that defines us, but the depth of our gratitude for the wonderful diversity that surrounds us. So, the next time we flash our pearl-white smile in front of the mirror, let us remember the beaver, the shark, the snail, and the platypus. In their smiles lies the true magic of nature, a reminder that in a world full of teeth of every shape and size, there is beauty to be found in every single smile.

12 PAIN TOLERANCE

Once upon a time, in a world where animals and humans coexisted without the looming threat of lawsuits, a peculiar debate emerged that captured the attention of the inhabitants of this shared world: the debate on pain tolerance. In this extraordinary world, the animal kingdom harbored a silent amusement, convinced that their ability to endure pain surpassed that of humans, who appeared to be melodramatic. Let us embark on an exploration of this absurdity, dissecting various aspects of pain tolerance and drawing parallels between human expressions and the stoic composure of their earthly counterparts.

Humans, with their extraordinary ability to emit anguished screams at the mere sight of a paper cut, have long boasted the sophistication of their vocal cords as finely tuned pain receptors. On the other hand, animals navigate gracefully through the wild, responding to their wounds with the subtlest of grumbles or a well-calibrated "Ouch!" The grandeur of human suffering, often expressed through elaborate vocalizations, becomes a source of amusement for creatures who communicate their pain in more reserved ways.

Consider, for a moment, the bold daredevils among humanity—the self-proclaimed adventurers who shudder at the mere sight of a needle. In stark contrast, the animal kingdom engages in high-risk activities daily, from evading predators to leaping across treacherous terrain. Their adrenaline-fueled endeavors make human apprehension toward a needle seem almost comical.

The kings and queens of human drama turn minor discomforts into full-blown soap operas. A stubbed toe becomes a tragedy of Shakespearean proportions, with anguished cries echoing in the air. Meanwhile, animals face life's adversities with a impassive attitude, or its equivalent, if they had lips.

A wolf would not shed tears over a thorn in its hoof; it stoically removes the obstacle and continues its majestic journey through the wild.

In the realm of pain, humans may boast of being the undisputed champions, with elaborate stories and theatrical expressions. Yet, animals, with their stoic demeanor, raise an eyebrow—or its animal equivalent—at this claim. Whether it's a cat landing on its feet after a fall or a dog shaking it off nonchalantly after a collision, animals embody the spirit of "Keep calm and carry on" with an elegance that surpasses most motivational posters.

Humans, in their pursuit of understanding and controlling pain, have created intricate psychological theories and therapeutic practices. In contrast, animals have mastered the art of self-soothing. Dogs, for instance, literally lick their wounds, offering the world an original form of ice cream therapy without the hefty costs associated with human therapy sessions.

As we conclude this contemplation, dear reader, imagine sitting there tending to a paper cut, meditating on the apparent injustice of life. In that moment, think also of the unsung heroes of pain tolerance—our earthly companions. The animal kingdom continues to endure life's trials with a grace and resoluteness that humanity can only aspire to reach. After all, who needs a pain threshold when you can conquer the world with a nonchalant tail flick?

Turning to another facet of this curious debate, consider the great exaggerators among humans—the undisputed champions of storytelling, who elevate the memory of their pain experiences to an art form. A simple paper cut transforms into an anguished saga of epic proportions, told with dramatic pauses and embellishments that rival even the most skilled bards. In this realm of human drama, animals communicate pain without the need for elaborate narratives. A wounded gazelle, for example, simply moves away from a predator, leaving the stage to the humans who insist on turning every minor scratch into a three-act drama.

While humans seek inner peace through yoga studies, animals have practiced the art of Zen for millennia. Consider the humble cat, whose ability to contort into impossible positions rivals even the most experienced yogis. Animals embody both physical and mental flexibility, adapting to discomfort without the need for a tranquil studio backdrop or a subscription fee.

Humans may boast of their ability to throw sly glances, but animals excel at this art. Imagine a squirrel losing its winter acorn stash. Does it throw a fit? No. Instead, it casts a scornful look at the universe, as if to say, “Really?” Animals face life’s disappointments with a nonchalant expression that even the most composed poker player would find uncomfortable.

The next facet of this exploration brings us to the unofficial proponents of DIY pain relief. Humans invest significant sums of money in pain relief products, from luxury creams to elaborate massage gadgets. On the other hand, animals embrace the concept of DIY pain relief. Observe a bird meticulously preening its feathers or a bear scratching its back against a tree. In their world, there’s no need for a spa day when nature offers the ultimate pain relief regimen.

In the human world, a simple headache can elicit a symphony of complaints and requests for sympathy. Animals, on the other hand, proceed through migraines of mythical proportions without uttering a single complaint. When was the last time you saw a stoic turtle endure a thundering headache? Probably never, because they’re too busy embodying the ancient philosophy of “smile and bear it.”

While humans boast about their sophisticated culinary experiences, animals emerge as the true epicureans of foraging. From deciphering the ripest fruits to locating the juiciest insects, animals navigate the world of gastronomy with a skill that would make any Master Chef envious. In their world, pain tolerance takes a backseat when a gourmet meal is at stake.

Humans may have created entire industries around stress relief and therapy, but animals have been offering purring therapy for centuries. The rhythmic hum of a satisfied cat has a calming effect that no mindfulness app can replicate. Perhaps humans could take a cue from their feline friends and embrace the therapeutic power of purrs instead of endlessly seeking the next miraculous stress-relief technique.

Humans may find comfort in the fact that, despite their tricks and elaborate rituals for pain, animals remain the unsung heroes of pain tolerance. So, the next time you stub your toe and reflect on the injustice of it all, remember that somewhere, a lethargic squirrel is giving you a scornful glance, silently questioning your commitment to the art of enduring discomfort.

Turning to another dimension of this exploration, consider the masters of transformation among humans, who pride themselves on their ability to thrive in new environments. However, animals have perfected the art of transformation since the dawn of time. From the depths of the ocean to the highest mountain peaks, animals acclimatize effortlessly to their surroundings. They endure extreme temperatures, navigate treacherous landscapes, and face formidable predators, all without the need for air-conditioned rooms or high-tech gear.

While humans scramble to tackle the environmental crisis they have largely created, animals stand as eco-warriors. They don't need climate change summits or green energy campaigns; they instinctively understand nature's delicate balance. Animals have lived sustainably for eons, demonstrating an innate wisdom that makes human attempts at environmental conservation seem like pale imitations.

Long before the term "social distancing" became a global buzzword, animals were experts at maintaining personal space. Whether it's a herd of elephants gracefully keeping a safe distance or a solitary cat finding comfort in a quiet corner, animals understand the importance of boundaries without the need for awkward conversations or six-foot distance markers.

In a world that never stops buzzing with human chatter, animals emerge as the poets of silence. They communicate through subtle gestures, nuanced expressions, and the eloquent language of body movements. While humans engage in endless debates, animals convey profound messages with a single glance or the gentle rustling of leaves. In the realm of connection, animals are the true masters, leaving humans scrambling for words.

Humans, with their reliance on patterns and architects, may think of themselves as the true masters of survival. However, animals have perfected the art of survival through instinct alone.

From building intricate nests to constructing elaborate burrows, animals create habitats that seamlessly blend with the natural world. Humans may build skyscrapers and marvel at their architectural achievements, but animals silently construct homes that harmonize with the Earth.

While humans boast of their technological advances and scientific discoveries, animals carry with them the wisdom of generations. From migration patterns encoded in the DNA of birds to the complex social structures of ant colonies, animals possess a wealth of knowledge that predates humanity's oldest civilizations.

So, while humans continue their pursuit of superiority, perhaps it's time to acknowledge that in the intricate dance of life, animals lead with a grace and wisdom that humanity can only aspire to emulate. After all, in the vast symphony of existence, every creature plays a unique note, and the harmony of Earth depends on the collective wisdom of its inhabitants.

Animals live by the ethic of "just enough," wisely using resources and ensuring waste is minimal. Humans, with their insatiable desire for more, might take a lesson from these ecological minimalists.

While humans boast of their technological advances and scientific discoveries, animals silently carry with them the wisdom of generations. From migration patterns encoded in the DNA of birds to the intricate social structures of ant colonies, animals possess a wealth of knowledge that predates humanity's oldest civilizations.

Shifting to another layer of this exploration, consider the timeless healers among humans who boast of their medical progress. Animals, however, possess an instinctive knowledge that transcends human medical advancements. Creatures both large and small have the innate ability to seek natural remedies, whether it's a wounded bird applying antimicrobial saliva to its wounds or a monkey using medicinal plants to heal the afflicted. Animals embody a healing wisdom that predates the era of pharmaceutical laboratories and complex surgical procedures. Perhaps humans could take a step back and learn from the original healers of the natural world.

In the realm of emotional intelligence, humans struggle with complexities, while animals navigate the intricate landscape of feelings effortlessly. From elephants mourning their dead to dogs sensing the emotions of their owners, animals display an emotional acuity that eclipses the most advanced human psychological theories. Without the need for therapy sessions or self-help books, animals embody a deep understanding of the emotional currents that shape their existence.

Humans, with their GPS systems and navigation apps, may feel they have mastered the art of orientation. However, animals have been cosmic navigators long before satellites adorned the skies. From the migratory patterns of birds to the celestial navigation abilities of sea turtles, animals traverse vast distances with a remarkable sense of direction. In the grand cosmic dance, animals move seamlessly through the rhythms of the Earth and stars, guided by an innate compass that remains a mystery to human science.

While humans create art galleries to showcase their interpretations of beauty, animals are the true aesthetes of nature's palette. From the vibrant plumage of tropical birds to the intricate patterns on a butterfly's wings, animals wear the colors of the Earth with unparalleled grace. Their natural adornments rival the most intricate human designs, proving that Mother Nature remains the ultimate artist. Humans may attempt to capture the essence of beauty on canvas, but animals live and breathe it every moment.

13 BEARD AND HAIR, THANK YOU!

Welcome back, dear readers, to our sardonic and unrestrained exploration of the human fascination with fur and hair. By now, we know that Homo Stultus is the strangest species on Earth. As we continue our journey through this bizarre realm, let's shed some light on the notion of human hair as a fashion statement.

In the animal kingdom, fur is a practical necessity, a developmental transformation that protects creatures from the elements. But, oh, the ingenious human mind takes this concept to new heights!

Humans transform the "fur" on their heads into a canvas of personal expression, using it to communicate everything from their mood to their rebellious attitudes. They dye their hair every imaginable color, from rainbow hues that catch the eye to the captivating silver of a weathered moon. They even invent whimsical names for these colors, like unicorn pastel, cosmic lavender, and mermaid turquoise, just to name a few.

However, the most extraordinary aspect of human hair fashion is wigs. Why settle for the natural strands with which you were born when you can have a fresh and stimulating mane every day of the week? Humans have an insatiable appetite for variety, and wigs offer a buffet of hair options that leave the animal kingdom in envy. Human hair is harvested, styled with the sole purpose of being worn by someone else. It's as if the human species collectively decided, "Why settle for what nature gave us when we can borrow someone else's locks?"

Wait, there's more! Humans' relationship with their facial fur goes beyond mere grooming; it becomes an embodiment of their personality. It is said that a man's beard symbolizes wisdom, sturdiness, and masculinity, while a woman's mustache-shaped eyebrows, an eccentric addition to their facial repertoire, reflect boldness and cutting-edge fashion.

Yet, despite all efforts to maintain a perfect appearance, there are times when humans suddenly embrace their inner wildness – the long-standing tradition of "No-Shave November" or "Movember" for men. During this magical month, razors tremble in fear as men everywhere let their facial follicles grow freely, proudly sporting their wild beards like badges of honor. Of course, this is only temporary, because in the following month, precision grooming and an abundance of beard oils will make their return.

And let's not forget the fascinating phenomenon of body hair removal in humans. It seems that the standardization of a hairless perfection is reserved for specific areas of the body, while others are curiously left untouched. Society decrees that the face, armpits, and legs must be as smooth as a marble sculpture, yet eyebrows are left to grow unchecked, like rebellious weeds, adding an element of casualness to their carefully curated appearances.

The beauty industry, recognizing the human desire for an eternal quest for perfection, has skillfully capitalized on these peculiar habits. Beauty salons, aesthetic centers, and laser clinics become temples of self-improvement, where humans invest time and money to remove, style, or replace their hair with almost religious fervor.

As we venture into the next chapter, we will explore further the fascinating world of human fashion – an exciting expedition where dressed creatures parade, trying to impress each other with their fabric-based audacity. Until then, remember to appreciate your unique fur and hair, for it is this very peculiarity that makes you a true Homo Stultus.

Are we sure that humans are the most evolved beings? Let's talk about our skin. We are like hairless monkeys who forgot to wax and shave properly. We have a bit of hair here and there, and if we're lucky, a bit of hair on our heads. That's it. On the other hand, if we look at the vast majority of animals, they have bodies covered in fur, feathers, scales, or shells. Often quite pleasing to the eye. Without going into the details of how we lost our fur but not the habit, which is not the purpose of this writing as it aims to be a contemplative amusement rather than a scientific treatise, we can easily understand how much we depend on external objects and materials for our bodies.

We can't walk around naked because, in addition to attracting unpleasant encounters with passersby and the police, we would freeze to death in the winter. We always need something to act as fur for us. The same goes for summer. Without protective membranes, we always feel hot even if we had a fan on our heads.

Instead, let's look at some animals. Let's start with our dogs and cats. Regardless of the species, more or less, they have the ability to shorten or lengthen their fur on their own. I know, you're thinking of the Chihuahua, but I said more or less, and some people don't even consider the Aztec dog a proper breed.

If we look at wild animals, besides primates who have hair everywhere except near their orifices (all orifices! Wouldn't it be cleaner down there?), we also see the fascinating mechanisms of birds. They have a faster metabolism and a higher body temperature than ours. This means that birds spend more energy keeping their bodies warm. Furthermore, in winter, they struggle more to find food. So, when the temperature suddenly drops, they literally need to seek shelter, such as caves and tree bark.

Starlings and sparrows settle for roofs or dense vegetation. If that's not enough, they fluff their feathers to trap heat. Some birds, like emperor penguins, use solidarity by huddling together. In extreme cases, some species of birds enter a sort of hibernation, lowering their body temperature by 10 or 15 degrees.

Additionally, there are the galliforms, like the white partridge and pheasants, which, in rather cold temperatures, bury themselves under the snow to conserve energy, like the Inuit in their igloos. The smartest, in my opinion, is the Siberian tit, which does nothing. It neither moves nor eats. This way, it doesn't waste energy.

Can you imagine a customer at the golden chain fast food who doesn't move or eat in winter? It would probably do them good, and they should have enough energy in their spherical belly.

However, the best and most common solution among birds is migration. It's cold? I'll go to the Caribbean. Simple. Indian geese, for example, fly over the Himalayas at over 8,000 meters, in sub-zero temperatures and strong winds. They expose themselves to long journeys and dangers, but the reward is warmth. Even Scottish cuckoos, when they get tired of the northern cold, head toward Africa. If they realize the travel conditions aren't favorable, they stop in the forests of southern England or turn back.

For us, Homines Stulti, it would all be either too easy psychologically or too difficult physically. Imagine our naked man at the fast food in an igloo :dead from the cold after a few hours of suffering. Or imagine all of us, dressed or not, it doesn't matter, suddenly stopping on the street, maybe at the bus stop, and huddling together to keep warm? It would be very easy, but the social norms we've set for ourselves wouldn't allow it.

Or let's talk about migration. Most of us do boring, unsatisfying work for eleven months and then only two weeks a year to migrate to a warmer, nicer place where we do nothing. Wouldn't it make more sense to do exactly the opposite? Like our holidays, two weeks in a terrible place and the rest of the year where we like it most? It's not possible, we have work, family, but above all, social pressure that never leaves us in peace. And then, the plane ticket is not exactly cheap.

Some of us, let's call them Homines Illuminati, manage to better control both the mind and the body, for example by increasing body temperature through regular breathing exercises and meditation, but if it's cold, it's cold!

In this light, the beauty industry emerges both as a facilitator and a reflection of the human pursuit of perfection. Salons and clinics offer a range of services aimed at shaping and enhancing natural features, catering to an audience eager to meet society's standards of attractiveness. However, beneath the surface lies a deeper desire for self-expression and validation, as individuals strive to reconcile their inner self with external expectations.

Delving deeper into the intricate dynamics of human hair, where every follicle becomes a thread in the fabric of self-expression, let's consider, for example, the cultural significance attributed to hairstyles throughout history. From the elaborate hairstyles of ancient civilizations to the sober and minimalist styles of modern times, hair has served as a reflection of social norms and values.

In ancient Egypt, for example, elaborate hairstyles adorned with jewelry and intricate braids were reserved for the nobility and royalty, serving as symbols of wealth and status. Meanwhile, in medieval Europe, hair was often hidden under veils and head coverings as a sign of modesty and piety. Fast forward to modern times, and hairstyles are imbued with an entirely new set of meanings, ranging from statements of rebellion to assertions of individuality.

Moreover, the act of styling and grooming one's hair has become a ritualistic practice, with entire industries dedicated to providing products and services aimed at achieving the perfect look. From shampoos and conditioners to gels and styling serums, the options are endless, suited to every imaginable hair type and texture. And let's not forget the countless hours spent in front of mirrors, meticulously arranging each strand into place, in the pursuit of the elusive beauty ideal.

But perhaps what is most fascinating is the cultural diversity that exists in the realm of hair. From the intricate braids of African tribes to the elaborate headdresses of indigenous peoples, hair serves as a canvas on which cultural identity is painted. Each style carries with it a rich history and tradition, passed down through generations, serving as a reminder of the interconnectedness of humanity.

14 MASTERS OF SURVIVAL

Another intriguing paradox: humans, boasting intelligence and dominance over the animal kingdom, often find themselves overshadowed by the remarkable abilities of our furry, feathered, and finned companions when it comes to recognizing and escaping danger. Nature, in its vast wisdom, seems to have granted the true survival skills to the seemingly modest inhabitants of our planet.

Thus, we embark on an extravagant expedition through the enchanted realms of the wild, where creatures we often consider "inferior" unveil their extraordinary talents for navigating the dangerous dance of survival with a mastery that leaves us humans in a perpetual state of awe. It is a spectacle that unfolds like a masterclass in cunning and instinct, forcing us to reassess our perceived supremacy in the grand theater of existence. We begin with the enchanting symphony of perception, where our fellow animals orchestrate a hypnotic harmony that makes the human approach little more than a cacophony.

For example, the common domestic cat possesses eyes that function like intricate radar, detecting the slightest movements with precision that makes our advanced security systems seem like child's play. A rustle in the underbrush, a flicker in the shadows—and there, the kitty is on high alert, ready to gracefully evade danger with the skill of a seasoned ninja.

In contrast, humans, absorbed by the brightness of screens and trapped by the clamor of our own creation, often find themselves clumsy when faced with the unexpected.

Our phones may be "smart," but it is the untamed creatures that hold the true keys to survival. The symphony extends beyond mere visual acuity.

In a world where technological wonders dominate our daily lives, it is easy to be mesmerized by the sleek design and seemingly endless capabilities of our smartphones and gadgets. These devices, hailed as the pinnacle of human ingenuity, have become indispensable tools for contact, navigation, and entertainment. However, amidst the clamor of our digitally driven existence, it is crucial to pause and reflect on the eternal wisdom possessed by the untamed creatures of the natural world. Nature is the ultimate teacher, offering lessons that go beyond the bounds of human invention. In an age of intelligent devices, capable of processing vast amounts of information in the blink of an eye, they lack the depth and complexity of ecosystems that have evolved over millions of years.

In the wild, survival is not just about downloading the right app or accessing the latest data; it is a multifaceted dance of interpretation, cooperation, and hardihood. Consider the complexities of a forest ecosystem, where every plant, animal, and microorganism plays a vital role in maintaining the delicate balance of life. From the majestic trees that provide shelter and sustenance to a myriad of creatures, to the modest fungi that recycle nutrients and enrich the soil, every organism contributes to the greater whole. In this interconnected web of life, there are no shortcuts or quick fixes, only the relentless pursuit of balance.

One of the deepest lessons nature teaches us is the importance of acclimatization. Unlike our smartphones, limited by the constraints of their programming, living organisms possess the remarkable ability to evolve in response to changing conditions. Through countless generations of trial and error, species have refined their survival strategies, developing specialized traits and behaviours that allow them to thrive in diverse environments. Take, for example, the cunning fox, whose acute senses and agile movements allow it to outsmart prey and evade predators in the harsh desert. Or consider the resilient desert plants, equipped with specialized customizations like succulent leaves and deep root systems that enable them to withstand scorching temperatures and prolonged droughts. These extraordinary feats of versatility remind us that survival is not a static state but an ongoing process of innovation and transformation.

In our fast-paced, technology-driven world, it is all too easy to lose touch with the rhythms of nature. We may find ourselves increasingly disconnected from the natural world, trapped in the digital distractions that permeate our daily lives. However, as we observe the untamed creatures that inhabit the wild places of the Earth, we are reminded of our place within the greater fabric of life. There is a certain humility in recognizing that, despite all our technological might, we are still bound by the same fundamental laws that govern the natural world. We may build towering skyscrapers and travel around the globe in a matter of hours, but we are ultimately bound by the cycles of the sun and moon, the ebb and flow of the tides, and the eternal dance of life and death. In the end, perhaps the true measure of our intelligence lies not in the sophistication of our gadgets, but in our ability to learn from the untamed beings who share this planet with us.

Animals are sensitive to the subtlest variations in the auditory landscape, deciphering the language of rustling leaves, distant footsteps, and the faintest echoes of danger. A symphony, rich and nuanced, in which each note is a piece of crucial information guiding them through the labyrinth of survival. Let us not underestimate the dormant power of instinct—a primal force flowing through the veins of our co-inhabitants. Animals, unburdened by the weight of excessive reflection and the paralysis of analysis, draw on ancient wisdom engraved in their DNA to navigate the perilous landscapes of survival. Consider the majestic salmon, performing navigational feats that make our GPS systems seem like relics of a bygone era. These extraordinary fish embark on epic journeys through oceans and rivers, guided solely by an innate sense of direction. Meanwhile, we humans, armed with our maps and compasses, often find ourselves disoriented in the concrete jungles we've built—a testament to the overestimated power of human intellect.

Instinct, that primordial whisper echoing through the millennia, guides animals in a dance with their environment. From the magnetic sense of birds guiding them across continents to the migratory patterns of butterflies, each species possesses an instinctive compass that puts our technological wonders to shame. No discussion of animals' extraordinary abilities would be complete without recognizing their unparalleled mastery of disguise. Enter the chameleon, that mystical wizard of the reptile kingdom, a master of camouflage with its environment—a skill that even surpasses the most sophisticated military camouflage. Consider the art of the octopus, a master of deception. With the ability to change color, texture, and pattern, the octopus transforms into a living kaleidoscope, evading predators and prey alike. Our attempts at blending in with our dull clothing and urban camouflage seem frankly comical in comparison.

When danger unfurls its sinister tentacles, animals reveal an escape choreography that rivals even the most thrilling action sequences. Imagine the cheetah, as fast as the wind, reaching speeds that make Usain Bolt, an eight-time Olympic gold medalist, and the world record holder for the 100 and 200 meters, universally hailed as the fastest person in history, seem rooted to the spot, or the seemingly modest hare performing gravity-defying leaps to evade predators. These maneuvers make our puerile attempts at escape—whether through sleek cars or high-speed chases—appear downright ridiculous.

While we find ourselves trapped in our metal boxes during rush hour, the sound of horns and the noise of shifting gears filling the air, animals perform seamless escapes with a mastery that mocks our feeble attempts at speed and agility. It's as if the animal kingdom has collectively conspired to teach their so-called superiors the art of eluding danger. Consider the humble earthworm, a seemingly defenseless creature, yet a master of underground evasion. When rain threatens, it senses the impending danger and digs deeper into the soil, escaping the perils of drowning. In our concrete jungles, we are forced to navigate flooded streets and traffic jams, our mighty machines rendered useless in the face of the instinctive wisdom of a simple earthworm.

As we continue our journey through the complex weave of life, we turn our gaze toward the marvelous intricacies of connection that permeate the natural world. link, in its many forms, serves as the lifeblood of ecosystems, facilitating cooperation, warning of danger, and creating bonds between individuals. Deep in the rainforest, where various sounds fill the air, creatures engage in a symphony of link that rivals even the most intricate human languages.

Here, the calls of birds echo in the canopy, signaling territorial boundaries and attracting potential mates. Meanwhile, primates exchange a rich repertoire of vocalizations, conveying emotions ranging from fear to affection with surprising clarity. But transmission extends beyond the realm of sound—it is a subtle dance of gestures, scents, and touch that binds social groups together. Consider the intricate courtship displays of the paradise birds, where males engage in elaborate dances and vibrant feather displays to win over potential mates. These rituals, steeped in symbolism and tradition, are a testament to the power of contact in the pursuit of reproductive success.

And what about the silent language of scents, where chemical signals convey a wealth of information imperceptible to the human nose? In the animal kingdom, smell plays a crucial role in marking territory, identifying kin, and locating potential mates.

Perhaps most fascinating of all is the phenomenon of mimicry, where animals evolve to imitate the appearance or behaviour of other species to survive. In the depths of the ocean, where camouflage is essential, creatures like the mimic octopus and the leafy sea dragon have perfected the art of blending into their environment, making themselves virtually invisible to both predators and prey alike. In the realm of deception, where survival often depends on the ability to outwit adversaries, animals employ a vast array of tactics to confuse and disorient their enemies. Consider the extraordinary case of the paradise bird, whose remarkable ability to mimic the sounds of its environment allows it to avoid attracting potential predators within striking distance. It is a testament to the power of adaptation and innovation in the perpetual arms race of survival.

But amidst the beauty of transmission lies the sad reality of extinction—a fate that looms over many species in the face of human-induced environmental changes. With the disappearance of habitats and the decline of ecosystems, the delicate balance of connection that sustains life on Earth is teetering. It is a solemn reminder of our collective responsibility to preserve and protect the natural world for future generations.

So, dear reader, as we ponder the wonders of contact that abound in the natural world, let us do so with reverence and respect for the intricate web of life that sustains us all. Because, in the end, it is our ability to listen, understand, and empathize with our fellow inhabitants of this planet that will determine our fate in the grand drama of existence.

Furthermore, consider the fascinating realm of insect communication, where chemical signals known as pheromones play a vital role in coordinating social behaviours and reproductive activities. Ant colonies, for example, rely on pheromones to communicate information about food sources, dangers, and the status of the colony. Through the release and detection of these chemical traces, ants can efficiently organize foraging expeditions, defend their territory, and regulate the division of labor within the colony. Similarly, bees use intricate dances known as the "tremble dance" to communicate the location of nectar-rich flowers to their hive mates. Through precise movements and specific vibrations, foraging bees can convey valuable spatial information, allowing their sisters to navigate directly to the most productive foraging sites.

Moving beyond the microcosm of insect societies, we explore the complex vocalizations of marine mammals such as whales and cetaceans. These highly intelligent creatures communicate through a vast repertoire of clicks, whistles, and pulsed calls, serving various social and ecological functions. For example, bottlenose dolphins use distinctive whistles to identify themselves and maintain social bonds within their group, much like a personalized name. These distinctive whistles are learned early in life and serve as a unique vocal signature, allowing individuals to recognize each other and communicate across vast oceanic distances. Furthermore, the hypnotic songs of humpback whales are thought to play a role in mate attraction and territory establishment during the breeding season. These elaborate vocal performances can travel miles underwater and are believed to convey information about the singer's size, strength, and reproductive status to potential mates.

Additionally, communication in the natural world goes beyond traditional sensory modes to include sophisticated forms of electrical signaling and bioluminescence. Electric fish, such as electric eels and electric rays, possess specialized organs capable of generating powerful electric fields, which they use for navigation, prey detection, and communication. By emitting distinct patterns of electric pulses, electric fish can transmit information about species identity, social status, and reproductive readiness to conspecifics. Similarly, bioluminescent organisms, such as fireflies and deep-sea creatures, produce light through biochemical reactions, which they use for mating attraction, camouflage, and predator deterrence. These fascinating light and color displays serve as a visual language, enabling organisms to communicate across vast expanses of darkness and depth.

Moreover, communication in the natural world is not confined to individual organisms alone; it also encompasses complex interactions within entire ecosystems. For example, the phenomenon of allelopathy, whereby plants release chemical compounds to inhibit the growth of nearby plants, can shape the composition and dynamics of plant communities.

Through chemical signaling, plants can compete for resources, discourage herbivores, and facilitate mutualistic relationships with beneficial microorganisms.

Similarly, the intricate network of interactions between pollinators and flowering plants relies on a delicate balance of visual, olfactory, and tactile signals to ensure successful reproduction. Through the coevolution of specialized traits and behaviours, pollinators and plants have forged intricate partnerships that underlie the functioning of terrestrial ecosystems worldwide.

Therefore, dear reader, let us embrace the cosmic game that is our supposed superiority over the animal kingdom. While we marvel at our own creations and enjoy the illusion of control, let us not forget to metaphorically tip our hats to the true masters of survival—the creatures that navigate the world with a mastery and grace that make our human endeavors laughably inadequate. Perhaps, in our relentless pursuit of progress and domination, we have overlooked the profound lessons that nature has desperately tried to teach us. The animals, with their modest splendor, stand as living, breathing reminders that the true essence of survival does not lie in the grandeur of our creations, but in the simplicity of our connection to the world around us.

15 A COMPLICATED TRACHEA

In this chapter, we embark on a delightful exploration of the trachea and the esophagus, those narrow tubes that enrich our existence with their proximity (and literally keep us on edge, as you'll soon discover). Let's start with the trachea, that slender air passage essential for our survival. For humans, it's almost as if Mother Nature played a trick on us. While animals have wider tracheas, we humans are given a truly narrow experience. Consider the giraffe, a creature blessed with a neck so long it defies nature's laws. Despite its impressive height, the giraffe's esophagus efficiently transports food from mouth to stomach.

This process is achieved through a combination of specialized muscles and a well-coordinated swallowing mechanism. In the animal kingdom, such modifications ensure that eating is a seamless and risk-free activity. Or think about the snake, a creature with a unique set of challenges due to its elongated, limbless body. Snakes have developed specialized structures to accommodate both breathing and the swallowing of large prey. Their tracheas, flexible and extended, showcase nature's adaptability in creating tailor-made solutions for specific lifestyles. Imagine this: the majestic lion, free on the savanna, roaring without a care in the world. Now, picture that same lion with a trachea as narrow as a straw, forced to produce a wheezy, feeble roar. Funny, right?

Yet here we are, the supposed rulers of the animal kingdom, with tracheas so narrow that we end up gasping and panting, contemplating the irony of our phylogenetic design. While our four-legged friends enjoy a relaxed process of ingestion, we humans are burdened with a balancing act every time we swallow. It's as if someone designed our esophagus to be the narrowest bridge between pleasure and the potential danger of choking. Well done, evolution, well done!

Some might argue that the closeness of the trachea and esophagus in humans is a testament to our versatility. "Why not make air and food share the same properties?" someone might quip, with a mischievous glint in their eye. Yet this seemingly efficient arrangement comes with a set of challenges. Picture a delicious meal suddenly turning into a dangerous adventure because of a poorly timed laugh or an unexpected sneeze. The trachea decides to join the party, resulting in a choking sensation that turns your dining room into a chaotic scene reminiscent of a horror film.

But let's not forget the true *pièce de résistance:* the epiglottis. This flap of tissue, whose main job is to prevent food from entering the trachea, is like a bouncer at an exclusive club. It's supposed to be the guardian, ensuring that only air takes the express lane to the lungs while food follows the scenic route to the stomach. Yet, in the grand tradition of human quirks, the epiglottis sometimes fails in its duties. A momentary lapse, an ill-timed sip, and suddenly you're caught in a battle between breath and bolus. It's a dangerous dance, and the stakes are high. If the epiglottis slips up, you might find yourself coughing and sputtering, desperately trying to regain your composure.

Other animals' tracheas and esophagi are masterpieces of efficiency, designed to facilitate the uninterrupted flow of air and nourishment. Take the elegant hummingbird, capable of rapid, prolonged flight while feeding on nectar with its specialized, elongated beak. The harmony between breathing and feeding in these creatures contrasts sharply with the complex and dangerous web of our own anatomy. As we reflect on the wonders of the human trachea and esophagus, let's take a moment to appreciate the delicate balance between breath and bite. In those narrow corridors, the comedy of our existence truly unfolds, reminding us of the peculiarities and complexities that make us uniquely human. Whether it's the challenge of breathing freely or the perilous ballet between food and air, our anatomy weaves a narrative that is both fascinating and, at times, humorously perplexing. Now, let's dive deeper into the legacy of survival tale of our trachea and esophagus.

The uniqueness of human respiratory and digestive systems lies not only in their functional complexities but also in the challenges they present. While the narrowness of our trachea might seem like a minor innovative flaw, it plays a crucial role in the complex symphony of human life. Let's consider the Darwinian pressures that shaped our ancestors. Early humans, with their upright posture, faced a different set of structural challenges compared to their four-legged counterparts. Our vertical stance required customizations in the anatomy of our respiratory and digestive systems. Although our narrow trachea can occasionally be a source of frustration, it serves as evidence of evolution's adaptive nature, accommodating the needs of bipedal locomotion. Furthermore, the close proximity between the trachea and esophagus can be seen as an efficient use of space within the human body.

This anatomical arrangement might have roots in the necessity of housing a larger brain, a hallmark of our species. In the intricate web of evolution, trade-offs are inevitable, and our unique anatomical layout is one of those compromises that has enabled us to develop complex cognitive abilities.

Returning to the animal kingdom, let’s explore the various adaptations that animals display in their respiratory and digestive systems. Whales, for instance, are masters of aquatic life, with respiratory adjustments that allow them to remain submerged for long periods. Their tracheas and esophagi are well-suited for underwater living, showcasing the remarkable diversity of anatomical solutions evolution has crafted. Birds, known for their intricate vocalizations and aerial prowess, possess respiratory systems adapted to high-energy activities. Unlike ours, their tracheas feature rigid rings that provide structural support, ensuring unobstructed airflow during both inhalation and exhalation. This setup enables the complex melodies of birdsong and the prolonged flight of species like the albatross.

Now, let's shift our focus to the esophagus and its role in the grand theater of digestion. Although the narrowness of our esophagus occasionally presents challenges, it can also be viewed through the lens of energy efficiency. In the wild, the ability to eat quickly and move on was often a survival advantage. A narrow esophagus facilitates rapid swallowing, allowing our ancestors to remain alert to potential threats and minimize the time spent in vulnerable positions while eating. The esophagus works in concert with other parts of the digestive system, like the stomach and intestines. Its narrow design helps maintain a smooth flow, leading to faster digestion overall.

Consider animals with specialized diets. The koala, for instance, has an esophagus highly adapted to process eucalyptus leaves. Its digestive system is tailored to break down the tough cellulose found in these leaves, illustrating the diversity of progressive solutions to dietary challenges. Returning to the theme of our esophagus, its narrow design also reflects the eating habits shaped by our ancestors. Early humans, as hunter-gatherers, faced uncertainty in food availability.

Let's look at other species. The anteater, with its long snout and specialized tongue, is perfectly adapted to consume ants and termites efficiently. The streamlined process of swallowing these tiny insects aligns with a survival pressures favoring assimilation for specific dietary niches. As we continue to uncover the mysteries of our anatomy, it becomes clear that the trachea and esophagus are not just conduits but key players in the drama of life. The challenges posed by their design reflect the complex genetic history that has shaped us. Occasional mishaps, like choking during a hearty laugh, remind us of the delicate balance these structures maintain.

Turning our attention back to the epiglottis, let's delve into its developmental significance. This flap of tissue, acting as a guardian between the trachea and esophagus, has likely undergone refinements throughout human evolution. Its role in preventing aspiration—when food or liquid enters the airway—has been crucial for our species' survival. In the animal kingdom, similar mechanisms exist to guard against aspiration. Marine mammals, such as seals, have protective reflexes that close their airways during dives, highlighting the convergent evolution of protective mechanisms in species facing similar challenges.

As we marvel at the interplay between breathing and swallowing in humans, it's essential to recognize the interconnectedness of our anatomy with the broader history of life on Earth. The quirks and challenges presented by our trachea and esophagus aren't isolated phenomena but are intricately woven into the rich tapestry of evolution. Swallowing, often taken for granted, becomes a nuanced process in the human narrative. Our esophagus, designed for efficient ingestion, reflects the adaptability of our ancestors to an uncertain environment. When we look once more at the animal kingdom, we find a multitude of dietary versatilities that parallel and contrast with our own. From the specialized beak of a hummingbird to the elongated snout of an anteater, each species has evolved to thrive in its ecological niche.

As we continue to navigate the quirks of our anatomical design, we embrace not only its oddities and challenges but also celebrate the suppleness and adaptability that make us uniquely human.

16 UNDERWATER

Ah, the majestic animal kingdom, where creatures of all shapes and sizes play in the water, blissfully unaware of the cumbersome limitations that afflict us as humans. Breathing underwater? Well, it seems our aquatic friends have outdone us here. You see, while we Homines Stulti are stuck gasping for air like asthmatic fish out of water, animals have mastered the art of underwater respiration. It's as if they belong to an exclusive gill club where humans are strictly banned, and the only entry requirement is to have scales or a sleek set of feathers.

Let's talk about fish, those aquatic marvels. They have gills that effortlessly filter oxygen from water, an elegant biological filtration system that makes our lungs look as clunky as outdated contraptions from a bygone era. While we scramble around with snorkels, regulators, and all sorts of diving gear, fish simply glide, breathing deeply with every fin stroke. How convenient!

Then there are the marine mammals, the show-offs of the underwater world. Delphinids, whales, and seals, with their lungs adapted to hold more oxygen, dive and surface gracefully, as if mocking our pathetic attempts at synchronized swimming. They can stay submerged for what seems like an eternity, making our pathetic free dives look like a child splashing around in a pool.

And let's not forget the turtles, those wise ancient beings that have been around for millions of years. They've perfected underwater breathing as an art form. With their sturdy shells and slow, deliberate movements, they make everything seem like a leisurely underwater picnic. Meanwhile, we lug around tanks and masks like aquatic porters.

And don't even get me started on amphibians. Frogs, with their charm, can hop between water and land without missing a beat. It's as if they have a secret amphibian handshake they refuse to share with us, leaving us sidelined, wondering why evolution dealt us such a lousy hand when it comes to aquatic skills.

So here we are, the supposed masters of the animal kingdom, left gasping on the shore while our aquatic counterparts frolic beneath the waves. Perhaps we should consider signing up for some underwater breathing lessons from our fish friends. Until then, we'll keep up our clumsy attempts at snorkeling and scuba diving, forever envious of the gilled elite.

As we wander along the sandy shores of envy, let's dive a

bit deeper into the marvelous world of those who breathe underwater. It's not just the gill wonders of fish or the lung masters of marine mammals that make us seem like a drowning comedy act; it's the sheer variety and efficiency that animals exhibit in their aquatic escapades.

Consider the humble crustaceans, those armored marine creatures that roam the ocean floor. While we're busy inflating bulky life vests and juggling weights to stay submerged, crabs and lobsters strut around with their exoskeletons, perfectly adapted to withstand the crushing pressures of the deep. They have the underwater fashion game locked down, while we're left adjusting our leaky goggles.

And let's not overlook the octopus, the underwater Houdini. This multi-limbed marvel doesn't need an oxygen tank or diving lessons; it can simply blend into its surroundings, vanishing like a master of underwater disguise. Meanwhile, we're left floundering around in our wetsuits, hoping to catch a glimpse of the mysterious world these cephalopod escape artists call home.

Yet, the true champions of underwater life are the microorganisms. Single-celled organisms like plankton and algae have silently dominated the underwater breathing game for eons. They don't need fancy scuba gear or oxygen tanks; they simply absorb sunlight and convert it into vital oxygen through photosynthesis. It's the original green technology.

Let's also appreciate the grandeur of coral reefs, those vibrant underwater cities built by tiny polyps. These architectural masterpieces provide homes and shelter for countless marine species, creating vibrant communities that make our suburban neighborhoods seem insignificant. We're left with our artificial reefs and sunken ships, trying to replicate a fraction of the biodiversity that coral reefs effortlessly host.

And we can't forget the deep-sea dwellers, those mysterious creatures thriving in crushing darkness where even the bravest of humans fear to venture. Bioluminescent fish and bizarre anglerfish with glowing lures navigate the abyss with an ancient, innate ease. They are perfectly adapted to a realm where our most advanced technology is rendered fragile and clumsy, a silent mockery of our feeble and fleeting intrusions into their world.

So, as we reflect on the aquatic prowess of the animal kingdom, it's clear that they have not only mastered the art of breathing underwater but have also created an entire ecosystem that outshines our aquatic efforts. Perhaps it's time for us to take a step back, acknowledge our underwater inadequacies, and tip our imaginary hats to the true champions of the deep blue. After all, in the grand tapestry of underwater life (sorry, I couldn't resist weaving in that word), animals have crafted a masterpiece while we're still struggling to thread the needle.

Now, as we dive deeper into the ocean of aquatic disparities, it's worth considering the progressive ingenuity displayed by creatures beneath the waves. While we, Homines Stulti, are busy devising complex contraptions to mimic the aquatic lifestyle, animals seem to chuckle at our feeble attempts from the comfort of their aquatic homes.

Take the jellyfish, those pulsing blobs of gelatinous grace. They drift through the ocean with an elegance that would put any synchronized swimmer to shame. Jellyfish let the water do the work. While we flail around with clumsy kicks, they serenely glide through the depths, the epitome of underwater cool.

Then there are the shrimp, underwater acrobats who have perfected the art of jet propulsion. As we desperately try not to kick our diving partners in the face, shrimp perform flawless underwater ballets with rapid tail flicks. It's as if they've attended underwater dance academies while we were stuck in land-based dance classes, forever out of sync with the rhythm of the sea.

Let’s also consider the magical realm of bioluminescence. Fireflies may light up our summer nights, but deep-sea organisms put on a light show that would make Broadway blush. Creatures like the anglerfish swing their personal lanterns to lure prey, a level of innovation that makes our underwater flashlight attempts seem embarrassingly mundane. It's as if they've attended the marine rave of luminescence while we're left with our glow sticks in the dark.

And the seahorse? That eccentric underwater equine? The male not only carries the eggs but also gives birth to adorable little seahorse babies. Meanwhile, human dads panic over diaper bags, a pale reflection of the paternal prowess demonstrated by these underwater equine fathers.

Let’s dive even deeper into the enigmatic world of deep-sea hydrothermal vents, where life thrives in the most unlikely of places.

Amidst boiling temperatures and crushing pressures, resilient organisms have adapted to harness energy from these otherworldly environments. From giant tube worms to bizarre bacteria, the inhabitants of these vent ecosystems defy our very understanding of life, reminding us that the ocean is a realm of infinite possibilities.

Then there are the migratory journeys of marine giants, like the majestic humpback whales that travel thousands of miles in search of food and mates. Their epic voyages span entire ocean basins, emphasizing the importance of global-scale conservation efforts. Throughout their travels, they face a multitude of challenges, from human-made obstacles such as pollution and climate change to natural predators and shifting ocean conditions.

Amidst these challenges, there is also robustness and hope. Conservation initiatives worldwide are working tirelessly to protect marine habitats and preserve the delicate balance of ocean ecosystems. From marine protected areas to community-driven conservation projects, individuals and organizations are coming together to ensure a sustainable future for our oceans and the creatures that inhabit them.

As we stand on the shores of this vast, mysterious realm, let's not forget the vital role each of us plays in shaping its future. Whether through simple acts like beach cleanups or advocating for policy changes on a global scale, we all have the power to make a difference. By embracing our connection to the ocean and fostering a sense of stewardship for its wonders, we can ensure that future generations inherit a world full of life and possibilities.

So, let's continue to explore, to learn, and to marvel at the wonders of the deep. Because in the ocean, we find not only a source of endless fascination but also a profound reminder of our collective responsibility to protect and preserve the natural world. For in the ocean's depths, we discover not only the secrets of the past but also the promise of a brighter, more sustainable future for all.

17 FLIGHT

Ah, the art of flight—a feat that has fascinated humanity for centuries, driving us to countless attempts to emulate the grace of our winged counterparts in the animal kingdom. As we delve into the disparities between our attempts at flight and the effortless mastery displayed by creatures with wings, our exploration is enriched by a myriad of examples from the vast animal world.

Wings, those exquisite pinnacles of avian splendor, are a testament to nature's unparalleled design. Birds, bats, and insects display these aerodynamic marvels with ease, leaving us humans marveling at their natural superiority. Meanwhile, with our ingenuity, we have opted to build colossal metal contraptions with roaring engines, bypassing the elegance of winged flight.

Who needs wings when you can have the deafening roar of

takeoff drowning out the poetry of bird flight? Imagine a bird taking flight, its wings unfolding majestically, catching the wind with a grace that makes our clumsy airplanes mere mechanical monsters. Birds don't need a 10,000-foot runway, a control tower, or a runway as long as an epic novel; they simply spread their wings and ascend into the boundless blue, leaving us fumbling with pre-flight checks and seatbelts.

Let's explore the spectacle of takeoff. Birds rise into the sky with an air of nonchalance, leaving the ground effortlessly. In contrast, the clumsy spectacle of humans, strapped to seats as they speed down the runway in a metal tube, is anything but elegant. The only "flight" happening here is the collective sighs of passengers hoping their in-flight meal doesn't cause gastric turbulence. Bravo, Humanity! You've turned a grand act into a high-stakes gamble on the runway.

Let's discuss freedom of movement—a concept animals grasp with unparalleled finesse, while we remain tethered to airport schedules and flight plans. Birds traverse continents on migratory journeys that make our globetrotting efforts seem trivial. Bats dive into the night, performing aerial acrobatics with precision, while we squabble over armrests and legroom at 30,000 feet. Insects, with their tiny wings, flutter with an agility that leaves our heavy airbuses in the dust.

As we continue to design flying machines that are faster, larger, and more efficient, let's take a moment to appreciate the true artisans of the air—birds with their melodic symphonies, bats with their nocturnal poetry, and insects with their intricate flight ballets. Perhaps there is a lesson to be learned from the effortless beauty with which they navigate the skies, a lesson that transcends the pathetic mechanisms and calculated precision of our airplanes.

Thus, the absence of innate flight ability in humans doesn't equate to stupidity. Our intelligence doesn't lie in our wings but in our minds—in our ability to think critically, solve complex problems, and shape the world around us. Although we cannot soar through the skies like our feathered friends, we have accomplished wonders far beyond the reach of flight, enriching our world in ways that are uniquely human.

Humans have long been fascinated by the ability to fly. From the ancient legends of Icarus to the modern marvels of aviation, flight has captured our imagination. We watch birds soaring in the sky and envy their freedom, their effortless navigation through the vast expanse of the atmosphere. Yet, despite all our scientific progress and technological achievements, we remain creatures tethered to the ground. Some might argue that this limitation is a testament to the shortcomings of human intelligence. It's true that, compared to many other species on this planet, humans have relatively limited physical strength.

Our bodies are not naturally equipped for flight. We lack wings, feathers, and the hollow bones necessary to achieve sustained movement in the air. But does the absence of flying abilities necessarily equate to stupidity? That's a harsh judgment. The human brain, though not optimized for aerial navigation, possesses an extraordinary capacity for cognition. We have developed complex languages, created intricate artistic masterpieces, composed symphonies that move the soul, and constructed towering structures that reach for the sky. We have harnessed fire, unraveled the mysteries of the universe, and connected the world through networks of information. Yet, it is precisely our intellectual achievements that have led to the perception of human inadequacy in the context of flight. We have built machines that can lift us into the air, defy gravity, and transport us across vast distances. And still, individually, we remain grounded and dependent on these technological advancements.

Why, then, have humans not evolved to fly naturally? The answer lies in the intricate balance between evolution and transformation. Throughout our phylogenetic history, flight was not a necessary trait for survival. Instead, our ancestors focused their cognitive efforts on developing other skills, such as tool-making, social cooperation, and problem-solving. These abilities gave us a competitive edge in our environment, leading to our species' dominance on the planet. Furthermore, the pursuit of flight requires a combination of factors beyond intelligence alone. It involves a complex interplay of biology, physics, engineering, and a profound understanding of aerodynamics. While humans have made significant strides in understanding these principles, the challenges of flight remain immense.

Consider the albatross, known for its extraordinary ability to glide effortlessly for thousands of miles without flapping its wings.

This magnificent bird stands as a testament to nature's ingenuity, demonstrating how aerodynamic principles can be harnessed for unparalleled efficiency in flight. Inspired by the albatross, engineers have developed advanced aircraft designs that mimic its graceful gliding movements, leading to more fuel-efficient and environmentally sustainable aviation technologies.

Similarly, think of the humble bumblebee, with its seemingly impractical wings and bulky body. Despite its unconventional appearance, the bumblebee is a master of flight, capable of maneuvering with agility and precision that surpasses many other insects. By studying the bumblebee's unique aerodynamics, researchers have gained valuable insights into the principles of lift and propulsion, paving the way for innovations in drone technology and the design of micro-aircraft.

Even the tiniest creatures, like the fruit fly, offer crucial lessons in flight mechanics. With its intricate wing movements and rapid changes in direction, it demonstrates a level of control and maneuverability that rivals the most advanced aerial vehicles. By analyzing the fruit fly's flight biomechanics, scientists have developed new algorithms and control systems for autonomous flying robots, unlocking new possibilities for applications in search and rescue, environmental monitoring, and agricultural surveillance.

More recently, the pioneering spirit of aviators like Amelia Earhart and Charles Lindbergh has captured the public's imagination and inspired generations to push the boundaries of what is possible in the realm of flight. Amelia Earhart, the first female aviator to fly solo across the Atlantic Ocean, became an icon of courage and determination, challenging the gender norms of her time and inspiring women worldwide to pursue their dreams. Likewise, Charles Lindbergh's historic solo flight across the Atlantic in 1927 captivated the world's attention and ushered in a new era of aviation. His bold feat showcased the potential of aircraft as a means of long-distance travel and paved the way for the development of global commercial aviation.

Beyond these iconic figures, the history of flight is also rich with tales of innovation, experimentation, and bold exploration. From the Wright brothers' first powered flight in 1903 to the moon landing in 1969 (yes, we went there… don't start…), humans have continuously pushed the boundaries of what is possible in aviation and space exploration. Today, the legacy of these pioneers lives on in the many men and women who continue to push the limits of aerospace engineering and exploration. From developing supersonic jets to exploring Mars and beyond, the quest for flight remains as vibrant and exciting as ever.

In summary, the pursuit of flight is not merely a scientific or technological endeavor; it is a testament to the indomitable spirit of human exploration and innovation. Whether inspired by myth, driven by ambition, or fueled by curiosity, humanity has always sought to defy gravity and reach for the skies. And while we may never truly fly with wings of wax and feathers, our quest for flight continues to inspire wonder, awe, and imagination in all who look up at the sky.

18 SHOCK WANTED

In the grand tapestry of existence, there is a profound dichotomy between humans and animals in their encounters with the captivating force known as electricity. Humans, with their advanced intelligence, have created a world intricately interwoven with wires, circuits, and power grids. Meanwhile, our humble animal companions navigate this electric landscape with a simplicity that mocks our supposed superiority.

Let us first reflect on the birds, those feathered philosophers of the sky, and their effortless dance with electricity. Unlike us, they don't worry about volts and amperes, nor do they need to attend elaborate safety briefings before perching on a high-voltage wire. The avian community, in its infinite wisdom, treats electricity as just another branch of their cosmic playground.

Now, dear readers, imagine the audacity of the human race. We, who proclaim ourselves masters of the universe, tremble at the thought of a simple electric shock. Humans wear insulating gloves and shoes, conduct thorough risk assessments, and fill their surroundings with warnings about the dangers of electric currents. Birds, on the other hand, couldn't care less. They cheerfully perch on high-voltage wires, a feathered orchestra of nonchalance playing in the midst of the symphony of electricity.

The dichotomy doesn't end with birds. Consider the aquatic wonders of the depths – fish, the uncelebrated poets of the underwater realms. Humans shudder at the idea of navigating through an electric field, but for fish, it's just another day in their liquid paradise. Some species have even evolved to detect and navigate using electric fields.

The electric eel, a marvel of aquatic ingenuity, generates electric shocks not for war or self-defense, but as a tool for communication, navigation, and hunting. In comparison, we humans, with our towering intellects, struggle to charge our smartphones without causing a small electrical disaster in our homes.

Amphibians, those humidity-loving philosophers, bask in the rain while humans run for shelter, fearing being struck by lightning. Frogs, with their bulging eyes and webbed feet, sing to the skies in the middle of a storm, embracing the very elements that make us flee for cover. How ironic that creatures seemingly so primitive have a more harmonious relationship with the electric symphony of nature.

Let us not forget the insects, those small titans of mettle. They flit around light sources, seemingly indifferent to the brilliance of artificial illumination. While humans hustle to invent the perfect lightbulb, insects have been performing aerial acrobatics around flames for eons, ignoring the mortal fear of being struck by a buzzing device.

In the grand scheme of evolution, animals have seamlessly adapted to the electric pulse of the planet, integrating it into their daily existence without the need for elaborate precautions and warnings. Humans, on the other hand, approach electricity with a mix of reverence and trepidation, an uncertain alliance between mastery and fear.

So, while you connect your devices, surrounding yourself with protections against power surges and treating electricity as a potentially lethal force, take a moment to reflect on the wisdom of the animal kingdom. Perhaps there is a lesson to be learned from the effortless grace with which birds navigate high-voltage cables and fish move through electric fields. In the grand tapestry of existence, we are but a thread, woven into the wires of our own creation.

Every time I see birds calmly sitting on high-voltage wires, staring at us with a defiant look – or at least that's how I imagine them – I often wonder why they can do it and we can't. Someone once told me that we could do it too if, hypothetically, we managed to hang from a wire with one hand, without using the other to close the electrical circuit and end up fried. The absurdity isn't in the fact that I should hang from an electrical wire, but let's suppose I did it; in that position, I certainly wouldn't be as comfortable as any bird. Since both of the bird's feet are on the wire, electricity doesn't pass through it. There is no circuit, its two feet are at the same electrical potential, and electricity travels along the wire instead of through the bird, so the bird isn't electrocuted.

This is probably what we were taught in school during physics lessons. But just as our math teacher lied to us when they said we wouldn't always have a calculator with us, our science teacher didn't tell us the whole truth either. The answer taught in school is simplified and only valid for a stationary system when the voltage doesn't change. If you've ever touched a car and gotten a shock, you'll know that in cases where the voltage changes, the answer is wrong. When a car is parked, there is no circuit because the car is isolated from the ground, but current can still flow through our bodies to the ground through our feet, causing a shock. The car has been charged to a high voltage, and when touched, it discharges through us to the ground. In this case, a circuit isn't needed for the charge to flow; it's the voltage change that occurs when coming into contact with the car that causes the shock. Electric lines carry alternating currents where electricity regularly changes direction. Over time, the voltage oscillates between positive and negative values, indicating the direction of the voltage flow. Voltage is the difference in charge between two points, while current, measured in amperes, is the speed at which the charge flows. Since the voltage in the wire changes constantly, the voltage on a bird sitting on a wire changes too. In other words, the bird is continuously charged and discharged through its legs.

Another thing to consider is the bird's ability to collect and store electric charge. Since birds are very small and have a spherical shape, they can't store much electric charge. In fact, when a bird perches on a 50 Hz power line, it can carry only a microampere-level current. So, even though the bird has current flowing through its body, it is too small to fry it or even give it an electric shock. For comparison, a current of around 10 milliamps is enough to produce a painful shock in humans, and electrocution is possible from around 50 milliamps. We don't know if birds sitting on power lines can feel the current flowing through their bodies. Maybe they enjoy the vibrating sensation that runs through them. Perhaps this explains why we don't often see large birds, with greater capacity, sitting on high-voltage wires.

Although most birds are safe when resting on high-voltage wires, tens of millions of birds die every year due to electricity. Many are not electrocuted but simply die from the impact of a collision. However, other birds die from electrocution, and this happens for two reasons. Transmission lines carry electricity at much higher voltages, ranging from 275 kV to 400 kV, and up to 800 kV in some parts of the world, compared to distribution lines.

A high-voltage line can be strong enough to ionize the surrounding air and create a level of voltage that increases as you approach. This can be seen when aerial line technicians get close to a high-voltage pole with a conductive rod, and sparks fly before contact. If a bird approaches a high-voltage wire, it will sense the electricity generated and should avoid landing on it. However, if a bird flies at high speed or cannot maneuver away from the transmission line in time, the voltage gradient can be large enough to cause a current to flow through the bird, resulting in electrocution, or a shock so severe that it causes the bird to fall to the ground and die from the impact.

We humans, with all our technologies and precautions, seem to be prisoners of our own invention, while our animal companions navigate effortlessly through the invisible currents of life. Fish that use electric fields for navigation, frogs that sing in the rain, insects that challenge artificial light—each of them shows us an alternative way, a mode of being that is not bound by fear and control.

In the grand symphony of existence, perhaps we should learn to dance with electricity as birds do, to flow with the current as fish do, to embrace the storm as frogs do.

We might discover that true wisdom does not lie in dominating nature, but in integrating with it, accepting its power, and finding our place within it.

Thus, as we connect to our devices and try to tame this force that both fascinates and intimidates us, let us remember that we are just a piece of a much larger mosaic. A thread that, like the birds on the wires, can find its harmony in the grand design of the universe, if only we stop fearing and begin to dance.

19 LET'S NOT CLIMB

In the grand circus of life, there's a peculiar show in which animals proudly display their superior climbing skills, leaving the *Homines Stulti* wondering if we're just dragging our knuckles up the natural selection ladder. Even though we've built skyscrapers and conquered mountains, let's take a moment to recognize the true kings and queens of ascension – the creatures that make scaling heights a walk in the park, or rather, a climb up a tree.

Let's start with the squirrels. These furry acrobats have perfected the art of climbing trees with such mastery that they make Spider-Man look like an amateur. As they leap effortlessly from branch to branch, one might be tempted to ask: "Why bother with elevators when you can just squirrel your way to the top?" The agility and grace with which these little rodents navigate the vertical world are enough to make Cirque du Soleil performers turn green with envy.

Let's also pay tribute to our feathered companions, the birds. While we struggle with bulky climbing gear and grueling mountain expeditions, these marvelous, winged beings simply spread their wings and soar into flight. Oh, how convenient it must be to avoid traffic and security checks by flying over everything! It's the birds' world, and we're just guests. Their ability to effortlessly rise above the daily challenges of life on land makes us question the sanity of our own terrestrial pursuits. Perhaps we should abandon our quest for mountain conquests and take up ornithology – after all, the true limit is not the peak of some snow-covered summit, but the sky.

And then there's the humble goat, a four-legged mountaineer who mocks our hiking boots and our trail snacks. With hooves that could challenge any rock climber, they navigate effortlessly across dangerous cliffs and dizzying peaks. Watching a goat scale a steep rock face is like witnessing a masterclass in mountain mobility. Perhaps we should take a few lessons from these daring, bearded creatures, who make scaling summits as easy as a stroll in the meadow.

Let's not forget the scaly animals, like geckos, who defy the laws of gravity with their feet (or whatever the lizard equivalent is). While we struggle with crampons and ice axes, geckos simply stroll up vertical surfaces as though it were a relaxing Sunday walk. Who needs to worry about altitude sickness when you can cling to walls like a reptilian superhero? Their secret lies in the marvels of the millions of hair-like structures on their feet, creating a molecular bond with surfaces that would make even the most experienced climber jealous. It’s the geckos' world, and we’re just trying not to slip off the rock.

As we contemplate the comical contrast between our climbing efforts and the effortless mastery of animals, it becomes clear that Mother Nature has bestowed some extraordinary talents upon her non-human children. Perhaps, in our quest to conquer heights, we’ve lost sight of the elegance that nature inherently possesses. Animals seem to mock our seriousness with their playful climbs and gravity-defying escapades.

With all our sophisticated equipment and intricate plans, are we perhaps overcomplicating the act of climbing? Should we reconsider our reliance on advanced technology, adopting a more organic and instinctive approach to climbing?

It’s as if we’ve forgotten the raw, intuitive connection with the environment that our animal counterparts seem to possess effortlessly. Perhaps it’s time for us to shed some of our technological baggage and embrace a more primal approach to climbing.

Imagine this: a group of climbers, inspired by the simplicity of nature, abandoning their high-tech gear and opting for a more natural ascent. Instead of meticulously planned routes, they trust their instincts and the guidance of the rocks beneath their fingers. No carabiners, no harnesses—just humans rediscovering the joy of pure climbing.

As the climbers make their way up the mountain, they channel the spirit of the mountain goat, approaching the challenge with intuitive grace. Their movements become a dance, a celebration of the symbiotic relationship between human ambition and the natural terrain. With every step, they reconnect with the primordial instincts that lie dormant within all of us, drawing upon the ancient wisdom of our animal ancestors. The cumbersome ropes and bulky equipment disappear, replaced by a sense of freedom and liberation. Like the birds flying above them, these climbers experience the thrill of unrestrained movement, unburdened by the weight of modern technology.

With every grip and foothold, they rediscover the thrill of tactile connection, feeling the texture of the rock beneath their fingers and the earth beneath their feet. And when they reach the summit, they are not greeted by the triumphant fanfare of conquest but by the quiet serenity of communion with nature.

Ultimately, the dilemma of climbing teaches them that true wisdom lies not in conquering nature, but in surrendering to it. It teaches them that the greatest achievements are not measured in height or distance, but in the depth of the connections we forge with the world around us.

And so, as they stand at the top of the mountain, gazing out over the vast expanse of the world below, they understand they are not alone. They are part of something greater than themselves—a cosmic dance of life and death, creation and destruction, unfolding since the dawn of time.

And in that moment, they are filled with a deep sense of gratitude—for the mountain, for the sky, for the earth, and for all the creatures that call it home. Because they know that, in the end, it is not the conquest of nature that brings true fulfillment, but the recognition of our place within it—as humble participants in the great circus of life.

20 FORGIVENESS

In the vast tapestry of memory, humans are often seen as those who carry the weight of grievances, unable to easily forget offenses and disputes. Contrary to the idea that people forget disagreements easily, our nature often leads us to hold on to memories of conflicts and friction, deeply influencing our future interactions. In this context, we can turn our gaze to the animal kingdom to explore how other species may be less inclined to hold grudges.

In the heart of the savannah, lions often fight for territorial control. A dominant lion may challenge the leader of another pride, but once the dispute is resolved, the animals quickly return to their daily routine. Lions do not hold long-lasting grudges, and although they remember boundaries and rivals, they do not let these memories dominate their existence. In contrast, humans can carry grudges for years. A quarrel between neighbors over property lines can leave an emotional scar that persists, affecting relationships and creating tensions that don't easily fade.

In the aquatic realm, dolphins are known for their complex social life and cooperative interactions. When disputes arise between them, such as over food or territory, odontocetes tend to resolve conflicts quickly and return to normal coexistence. They do not hold grudges and prefer to maintain harmony within the group. Humans, however, often hold onto the memory of wrongs. A disagreement between friends can leave lasting marks, with negative emotions resurfacing each time the event is recalled, making it difficult to return to a fully harmonious relationship.

Territorial disputes among birds, such as between eagles or sparrows, are settled with acts of strength and skill, but once the contest is over, the birds tend to forget quickly and move on with their lives. They do not carry the weight of past disputes. Humans, on the other hand, may cling to memories of arguments, allowing them to influence their emotions and future actions. A quarrel over a parking spot can be remembered for a long time, fueling negative feelings that may resurface in every subsequent encounter.

In the insect world, ants engage in fierce battles to protect their colonies and resources. However, once the conflict is resolved, they immediately return to their collaborative routine without holding grudges. For humans, even disputes over small matters, like who should run the next errand, can leave deep traces. Negative emotions associated with these discussions can repeatedly resurface, affecting cooperation and group dynamics.

Dogs, our loyal companions, show a surprising ability to let go of small grievances. If scolded, they may feel momentarily hurt, but they quickly return to showing affection and loyalty. Humans, in contrast, often remember wrongs for a long time. A small disagreement with a partner or friend may stay etched in memory, influencing how we view and interact with that person in the future.

Octopuses are known for their intelligence and adaptability. When they compete for resources such as food or shelter, a brief dispute may occur, but once resolved, octopuses do not cling to the memory of the struggle. For humans, even small disputes can leave a lasting trace. An argument over who should do the cleaning can become a source of resentment that lingers long after the conflict itself.

Elephants are no different. They are often celebrated for their impressive memory and strong emotional intelligence. However, even they, after resolving a dispute over water or territory, do not hold grudges and can return to living in harmony with the rest of the herd.

Let's look to the sky. In the colorful world of the birds of paradise, the males display vibrant plumage and intricate dances to attract the females. A rejection during these displays leaves no lasting marks; the birds continue to live and perform without the previous rejection affecting their future actions. Humans, on the other hand, may suffer deeply from a romantic rejection, carrying the weight of the negative experience and allowing it to influence their future romantic interactions.

Ultimately, while the animal kingdom demonstrates an impressive ability to let go of grudges and continue living without lasting resentment, humans often find themselves carrying the weight of past conflicts. Our selective memory allows us to overcome many difficulties, but at the same time, it makes us vulnerable to holding on to grudges and resentments.

As we observe nature, we can learn the art of letting go and living in the present, emulating the ability of animals to quickly resolve conflicts and return to a state of harmony. In this way, we can strive to free our minds from the weight of old disputes, navigating with greater ease through the complexity of human relationships.

21 LONGEVITY

In the vast spectacle of life, where the performers embody different shapes and sizes, we find ourselves contemplating the extraordinary display of longevity. Humans, often boasting of their supposed dominance in the marathon of existence, should not overlook the fact that animals are often the overlooked marathoners, moving gracefully through the journey while humans struggle in a frantic race. Consider the majestic turtle. This creature, with its tranquil approach to life, takes the concept of "slow and steady wins the race" to an entirely new level. Unconcerned with the hustle of human life, the turtle casually carries the weight of wisdom on its ancient shell. While humans anxiously chase the elusive fountain of youth, the turtle meanders, a living manifestation of the art of facing life at a measured pace.

Humans dedicate themselves to eating salads and sipping dubious green smoothies in an attempt to cheat time, while the turtle calmly eats lettuce without a care. It's as though they've discovered that the secret to immortality lies in a diet of vegetables and a peaceful stroll through life. And what about the lobster? These ocean wonders disdain the mortal battle against aging. While humans worry about anti-aging creams and invest in dubious elixirs promising eternal youth, the lobster proudly displays its age on its exoskeleton. The lobster's molting process, a natural act of shedding its outer layer and emerging anew, makes the concept of aging practically irrelevant. Yes, the lobster grows, molts, and practically tosses aging aside like yesterday's trash. Meanwhile, humans wonder if they will ever find the fountain of youth or if it's merely a myth perpetuated by clever tour guides.

Let's not forget the ever-elegant Turritopsis dohrnii, the so-called "immortal jellyfish." These gelatinous marvels have mastered the art of biological renewal, a skill that would make any Hollywood celebrity envious. When faced with starvation or physical threat, they can revert their cells to a youthful polyp stage, effectively restarting their life cycle (Bavestrello et al., 1992).[9]. While humans desperately cling to the hope that science will unlock the secrets of eternal youth, this jellyfish gracefully dances through the depths of the ocean, living its best life with not a single wrinkle in sight. Its

---

[9] Bavestrello, G., Sommer, C., & Sarà, M. (1992). Bi-directional conversion in *Turritopsis nutricula*. In *Aspects of Hydrozoan Biology* (pp. 137-140).

ability to cheat death by returning to a juvenile state is a metaphorical exclamation point on our human obsession with appearing younger. In our quest for longevity, it seems that humans are running in circles, chasing fleeting dreams, while animals like the Turritopsis dohrnii effortlessly dance through the ages, their biological defiance a subject of scientific wonder (Rich, 2012).[10] Their carefree grace mocks our desperate attempts. Perhaps we should take a cue from our animal counterparts and learn to embrace the rhythm of life, savoring each moment instead of racing toward an elusive finish line.

So, turtles, lobsters, and jellyfish of the world – the true masters of longevity, gracefully navigating through time while we humans stagger like clowns in a perpetual circus of our own making. As we continue our frantic search for the elusive elixir, let's not forget the comic irony of our race against creatures that have long understood that the secret to a long life is not in the race, but in the timeless dance of existence. Moreover, let's not overlook the humble elephant, the gentle giant of the savanna. While humans exhaust themselves with elaborate skincare routines and surgeries in a desperate attempt to erase the marks of time, elephants wear their wrinkles like badges of honor. Each line on their majestic faces tells a story of wisdom and experience, a narrative etched into the very essence of their being. Elephants, it seems, have embraced the

---

[10] Rich, N. (2012, November 28). Can a Jellyfish Unlock the Secret of Immortality? *The New York Times Magazine*.

beauty of aging without the need for expensive creams or the sting of Botox injections.

And what about the enigmatic naked mole rat? This creature, with its lack of fur and wrinkled appearance, defies conventional standards of beauty. While humans invest countless hours and dollars in anti-aging products, the naked mole rat thrives in its unconventional charm. It scorns the idea of conforming to social norms, proudly displaying its unique features as a testament to resilience and adaptability. Perhaps, in our search for eternal youth, we could take a lesson from the naked mole rat and celebrate our individual traits instead of succumbing to the pressure of societal expectations.

As we explore the many ways in which animals outpace us in the longevity game, it's worth recognizing the pluck of the humble sea urchin. These modest creatures, with their spiny exteriors and simple lives, have unlocked the code of biological immortality. While humans fret over the inevitability of aging and the ticking of the clock, sea urchins defy the rules of mortality by regenerating their cells indefinitely. It's an accomplishment that eludes our most sophisticated scientific efforts, leaving us in awe of the seemingly modest sea urchin's ability to navigate the currents of time with ease. In the grand panorama of life – oh, forgive the unintentional use of the forbidden word – we appreciate the diversity of strategies adopted by our fellow inhabitants on this cosmic stage.

As we marvel at the wonders of the animal kingdom, let's not forget the resilient Greenland shark, a creature that transcends traditional rules of aging. These ancient mariners, with their slow growth rates and incredible lifespans, challenge our understanding of the biological clock. While humans struggle with the inevitability of birthdays and the social pressure to remain eternally youthful, Greenland sharks navigate the frozen depths with a stoic indifference to the passage of time. The depths of the ocean, it seems, hold the keys to timeless wisdom, with creatures like the immortal jellyfish and the mysterious yet enduring Greenland shark challenging our very notions of mortality. These aquatic sages silently swim through the annals of time, leaving us terrestrials to ponder the mysteries of existence.

We also recognize the resilient spirit of the Bristlecone pine, a tree that stands as a silent witness to millennia of time. These ancient sentinels of the mountain landscapes endure harsh climates and extreme conditions, with some specimens boasting an age that spans thousands of years. As we weave through the rich narrative of nature's longevity, it becomes clear that the spectacle of life extends beyond the animal kingdom.

The ancient art of bonsai, with its meticulous cultivation of miniature trees, whispers a different tale of time and patience. Bonsai artists, through their careful pruning and nurturing, coax nature into a harmonious dance of perpetual youth within a confined space. In the miniature world of bonsai, the passage of time is not marked by the wrinkles of bark or the greying of leaves, but by the artisanal shaping of branches and the attentive manipulation of the growing substrate. Each bonsai tree, a living sculpture frozen in time, encapsulates the essence of longevity within a confined space. The art of bonsai invites us to appreciate the beauty of life within the constraints of our circumstances, reminding us that the pursuit of eternal youth can be found in the crafted cultivation of each fleeting moment.

As we reflect on the various manifestations of longevity, both in the natural world and in humanity's artistic endeavors, it becomes evident that the grand circus of life is a multidimensional performance. From the slow and steady pace of the turtle to the ageless wisdom of ancient trees, from the mysterious depths of the ocean to the microscopic grit of bacteria, and from the enduring art of bonsai to the enigmatic stories of clams and whales, every act in this cosmic performance contributes to the overarching narrative of existence.

22 HABITAT DESTRUCTION

Humans, who have self-proclaimed themselves as masters of the planet, have managed to create a masterpiece of destruction like no other in the name of development. Let's take a moment to appreciate the art of transforming flourishing ecosystems into ever-expanding urban landscapes. It's a true tribute to human ingenuity: who needs forests, wetlands, and pristine landscapes when you can have towering skyscrapers, airports, and parking lots? The animals must thank their lucky stars for such generous gifts! The delicate balance of nature, carefully shaped over millions of years, has been disrupted by humanity's impressive endeavors.

It's as if the animals were just sitting there waiting for humans to arrive and showcase their skill in habitat destruction. What a fantastic era it must be to be alive for the unsuspecting wildlife, forced to adapt to futuristic designs of concrete and steel. In the animal kingdom, it's common knowledge that the pinnacle of success lies in how quickly one adapts to the changing environment.

Take, for example, the magnificent art of deforestation, a human specialty that has given animals a front-row seat to witness the majestic felling of their homes. Who wouldn't want to trade their cozy nests for the thrill of dodging bulldozers and logging trucks? And let's not forget the exquisite touch of pollution, a refined blend of chemicals and waste that adds that certain je ne sais quoi to the air and water. The animals, with their limited understanding of life's finer things, may not fully appreciate the aesthetic appeal of an oil spill or the gray smog-filled sky. But rest assured, they are undoubtedly relishing the opportunity to evolve and develop gills to survive in this newly improved aquatic desert. As humans continue to expand their urban jungles, animals everywhere must be overwhelmed with gratitude.

No one would refuse to exchange their natural habitats for the privilege of navigating through traffic and avoiding the occasional collision with a speeding car? It's a small price to pay for the chance to participate in this grand experiment of anthropogenic evolution. To sum up, dear reader, let's raise a glass to humans, the true architects of change. Through their innovative techniques of habitat destruction, they have generously invited animals to partake in the excitement of an urban safari. Who needs the wilderness when you can have the curated chaos of human progress? Cheers to the human touch – the touch that leaves no habitat intact!

On the other hand, let's see what some animals do. Let's observe the beaver, that master engineer of the animal kingdom. While humans boldly wield chainsaws and bulldozers to reshape their surroundings, beavers, in stark contrast, have chosen a path of admirable moderation. Despite having teeth that could rival any sharp tool produced by humans, these industrious rodents somehow resist the urge to level entire forests for their dam-building efforts. One must applaud their restraint as they carefully sculpt their aquatic shelters without triggering a catastrophic wave of destruction right at their doorstep.

Let's consider the attitude of the humble squirrel, a creature often overlooked in the grand narrative of environmental transformation. Faced with uncontrolled human development, squirrels have opted for a more sensible approach. Instead of cutting down trees to build high-rise condos for the squirrel elite, these furry architects discreetly construct nests within the existing vegetation. They seem to understand the old adage: "If it's not broken, don't fix it." So, while skyscrapers rise and fall in human cities, squirrels thrive in their arboreal sanctuaries, untouched by the reckless ambition of urban expansion.

Birds, those feathered guardians of the sky, on the other hand, show remarkable dedication to historical preservation. Unlike humans who demolish ancient monuments to make way for parking lots and shopping malls, birds maintain an unconditional reverence for their nesting sites. Generation after generation, these winged architects renovate and reuse the same nests, ensuring the integrity of their historical homes remains intact. If only humans could grasp the wisdom of building on the foundations of the past instead of erasing them in search of the next big novelty.

Even the earthworm, with its seemingly simple existence, brings valuable lessons for the bold Homo Stultus. While humans engage in large-scale deforestation and soil degradation, earthworms tirelessly work to improve the ground beneath them. These humble creatures aerate the soil, promoting healthy plant growth without resorting to the destruction of vast green expanses. It seems that earthworms understand the importance of sustainability, a concept that, unfortunately, has escaped some of their more destructive land companions.

If we observe the vast realm of the nobility of insects, the monarch butterfly reigns not only for its beauty but also for its mysterious migratory patterns. These delicate creatures travel thousands of miles across North America, defying the odds of urban expansion and human-induced environmental changes. Rather than yielding to the assault of concrete, monarch butterflies show the tenacity to adapt, finding new routes and green spaces amid the expanding urban frontier. The annual spectacle of their migration serves as a reminder that even the smallest creatures can face the challenges of human progress.

And what about chameleons? Masters of camouflage that change color, they have found innovative ways to adapt to the concrete palette of urban landscapes. In the wild, these reptiles blend perfectly with their surroundings, using their mimicry skills to escape predators. In urban environments, they've raised the bar, showing the chameleon's ability to blend into a man-made environment. These adaptable creatures effortlessly navigate the concrete and steel jungle, using their natural talents to merge into the human-built surroundings. It is a testament to the chameleon's survival instinct, proving that even in the face of rapid urbanization, nature has its ways of thriving.

Raccoons, those cunning bandits of the night, have discovered how to turn the chaos created by humans into an advantage. Urban areas provide these intelligent creatures with an abundance of food sources, from trash bins to discarded fast food leftovers. Unlike their rural counterparts, urban raccoons have adapted to the quirks of human civilization, thriving in the concrete labyrinth. Their ingenuity in navigating urban challenges demonstrates an unexpected harmony between human development and wildlife, as raccoons cleverly exploit the surplus of human excess.

In cities, we also have city pigeons, often regarded as simple pests, who have become unlikely ambassadors of adaptability in the urban context. Amid towering skyscrapers and crowded streets, pigeons have found their own refuges — the ledges, rooftops, and nooks of man-made structures. Unfazed by the chaos below, they coexist with the city dwellers, creating a symbiotic relationship based on the availability of food scraps and shelter provided by the urban landscape. Pigeons, with their ability to thrive in the heart of urban chaos, challenge the idea that wildlife and human development are mutually exclusive.

Even in the heart of expanding cities, foxes have managed to carve out their own space. These clever creatures have adapted to urban life, using parks, gardens, and even abandoned buildings as their hunting grounds. Urban foxes demonstrate a remarkable ability to navigate the challenges of traffic, noise, and artificial lighting, showing that the boundary between wilderness and urban landscape is not as rigid as one might think. Their ability to coexist with humans while maintaining the appearance of their wild instincts highlights nature's adaptability in the face of urban expansion. In the animal kingdom, it's clear that not all species view human progress as an existential threat. Instead, they demonstrate remarkable adaptability, finding ways to thrive amidst the concrete and steel structures that dominate the modern landscape.

Also, in the realm of insects, fireflies light up the night with their enchanting displays. Urban development and artificial lighting may have obscured the natural brilliance of the night sky, but fireflies have adapted to this new reality. They continue to illuminate the urban twilight, creating magical moments in the midst of the urban landscape. Fireflies, with their ability to persist in the face of light pollution, remind us that even amidst human-induced changes, nature finds a way to preserve its intrinsic beauty. Unfortunately, there has been a dramatic global decline in fireflies, or "lightning bugs" as we call them, partly due to light pollution, pesticides, etc., and partly because most of female fireflies cannot fly, making it difficult for them to easily form new nests when disturbed.

Other urban animals, such as crows, renowned for their intelligence, have seamlessly integrated into urban environments, proving that avian brilliance knows no bounds. These feathered scholars have learned to navigate traffic, use tools to their advantage, and even adapt their calls to communicate in the chaotic urban landscape. Urban crows are a testament to the adaptability and problem-solving abilities inherent in the animal kingdom. Instead of succumbing to the challenges posed by human development, they have embraced the opportunities presented by the concrete jungle.

There are also some unusual aquatic residents: urban turtles. In the midst of urbanization, certain species of turtles have discovered unconventional habitats within city limits. Rainwater ponds, artificial lakes, and even drainage ditches have become unexpected homes for these shelled residents. Urban turtles exemplify the wildlife's ability to find refuge in seemingly inhospitable environments. Their presence in man-made aquatic features challenges the notion that only pristine natural habitats can sustain diverse ecosystems, showing the adaptability of some species to urban water bodies. However, the characteristics of these environments do not offer the same rich array of microorganisms and plants that maintain the health of natural rivers and lakes. Consequently, the reproductive capabilities of turtles are compromised, as is the quality of their life.

Another fascinating animal, the coyote, once primarily associated with rural landscapes, has successfully infiltrated urban territories. These adaptable canids have navigated through the concrete jungle, using green spaces, parks, and even abandoned lots as their hunting grounds. Urban coyotes, with their ability to coexist with human communities, demonstrate a harmonious fusion of wildlife and city life. Their presence challenges preconceived notions about the boundaries between wilderness and urban development, highlighting nature's resilient spirit in the face of human progress.

While humans may have altered landscapes and disturbed ecosystems, animals have shown an extraordinary ability to coexist, adapt, and even thrive amidst urbanization. Perhaps there is a valuable lesson to be learned from the intricate dance between nature and development, urging us to consider more sustainable and harmonious approaches in shaping the world we share with our fellow beings. Here's to the remarkable diversity of life, finding its place in the intricate mosaic of the modern world.

Years ago, I went to Tibet. A pleasant trip, if not for the average altitude of 4900 meters. I was among the lucky few who, armed with rhodiola rosea, a medicinal plant that helps with weakness, could tolerate the thin air quite well. As I explored those beautiful plateaus, I realized that the only ones struggling to breathe were us, the species Homo Stultus. Other animals hopped joyfully along the mountain slopes or relaxed on the vast meadows. We adapt the environment to our needs, often destroying and polluting it, while most animals adapt to the surrounding environment. Take the chamois, for instance, which lives at such high altitudes where the low oxygen concentration makes survival difficult. It has larger lungs, allowing for superior respiratory capacity, and many more red blood cells in its blood.

Bedbugs are another example of environmental assimilation. Two decades ago, they were rare, but they adapted to insecticides and have become widespread around the world. There is a specific reason for this: as we modify the environment through industrialization, our habits, and the effects of climate change, the animals and plants that live in cities evolve to adapt to these changes. In some large metropolises, such as New York, Chicago, or Los Angeles, mutated flies have been identified: they have developed the ability to live in underground areas, no longer needing constant blood meals, becoming more resilient, and thriving even in winter. Speaking of metropolises, blackbirds, peregrine falcons, and Jamaican kestrels are now urban animals.

Take the blackbird, for example. It has little in common with its ancestor. Now it has a longer beak, a higher-pitched voice to be heard above the car horns, and it doesn't like to migrate. Even spiders are becoming more accustomed to the city. Just look at the webs between streetlights, which attract all kinds of insects, like moths, which in turn have become more resistant to light bulbs. Not to mention a species of lizard in Puerto Rico that has adapted over the years, modifying its legs for better grip on the scorching asphalt.

And the polar bear? It has small ears to reduce heat, thick skin for the cold and camouflage, and large paws to walk effortlessly in the snow. Homo Stultus, on the other hand, always needs to dress heavily, and if it were to cheerfully walk in the snow, besides fatigue, it would be an easy meal for the bear. Not to mention the ears: too large for both the winter climate and the tropical one. I remember when a butterfly flew into my brother's ear. What pain, but also what absurdity! It can happen to any Homo Stultus.

In a climate completely different from the Arctic, we find another animal that apparently has different characteristics but whose adaptive traits are similar to those of the polar bear: the camel. It has large legs to reduce pressure on the sand, a hump containing fat to convert into energy, but that fat doesn't insulate its body because it doesn't live in the cold. Furthermore, its nostrils can close to prevent sand from entering the nose. Its eyelashes are also thick for the same reason. Finally, its lips are very thick to eat desert plants without being pricked by cactus spines. The same thickness for the mouth is also found in the giraffe.

Are we sure it's a good idea to continue in this destructive direction? Perhaps many of them will be resilient. What about us?

23 ADAPTING TO THE ENVIRONMENT

In our relentless pursuit of knowledge, we often overlook the innate wisdom found in the animal kingdom. As we grapple with climate change, it's worth considering the environmental wisdom inherently exhibited by our animal counterparts. Animals seamlessly adapt to changes in weather patterns, adjusting their behaviours, migration routes, and even reproductive cycles in response to the ever-changing environment. Contrast this with human efforts to combat climate change, often involving intricate negotiations, international agreements, and bureaucracy. While humans gather in conference rooms to discuss environmental policies, other animals, without the luxury of such deliberations, instinctively understand the rhythms of nature, demonstrating that perhaps the most effective solutions are the simplest.

Animals perform a mesmerizing display, a constant harmony with the ever-shifting rhythms of the environment. Each species, from the smallest insects to the mightiest mammals, has perfected its movements to navigate the challenges posed by climatic variations. The Arctic fox, a master of adaptation, embodies this with a coat that changes shades with the seasons, blending seamlessly into the snow-covered landscape as winter arrives. Nature's choreography extends beyond individual modifications to intricate group performances. For example, the synchronized migration of wildebeests across the Serengeti in Tanzania, a spectacle that unfolds with a precision that rivals the best orchestras.

These annual migrations are not guided by climate models or weather forecasts but are an ancient instinct embedded in the collective memory of generations, passed down through the millennia. As Earth's climate undergoes unprecedented changes, humans find themselves struggling with the consequences of their actions. The animal kingdom, in its silent wisdom, invites us to join the dance of climatic harmony. It is an invitation to adapt not only as individuals but as a collective, recognizing that the survival of a species is intimately connected to the well-being of the entire ecosystem.

In the age of disposable convenience, animals embody the art of sustainability. The grim dance of predator and prey maintains ecological balance, preventing overcrowding and ensuring the health of ecosystems. Animals waste nothing, using every part of their prey for sustenance and survival. In contrast, human consumption patterns, marked by excess and a lack of consideration for consequences, paint a stark contrast. Landfills overflow with discarded gadgets, single-use plastics, and the remnants of our throwaway culture. In the animal kingdom, there is no concept of waste, only a harmonious cycle of life and death that sustains nature's delicate balance.

Animals possess an intuitive understanding of medicinal plants, relying on nature's pharmacy to address ailments. From primates using specific leaves to ease stomach pains, to dolphins employing certain types of sponges to protect their skin, the animal kingdom demonstrates a deep connection to the healing properties of the Earth. Humans, in their pursuit of pharmaceutical advancements, often overlook the treasures bestowed by the natural world. The reliance on synthetic compounds, with their numerous side effects, stands in stark contrast to the holistic approach embraced by our animal counterparts. Perhaps, in our search for progress, we have inadvertently strayed from the very source of our existence.

In the complexity of survival, animals recognize the importance of play. From otters sliding along riverbanks for pure enjoyment to crows engaging in aerial acrobatics, the animal kingdom acknowledges the therapeutic value of joy. Humans, on the other hand, often perceive leisure time as a luxury rather than an essential component of well-being. The relentless pursuit of productivity has led to a society where play is relegated to the margins of adult life. Animals, in their seemingly frivolous antics, remind us that the art of play is a testament to a life well-lived, a lesson that, in our complexity, we would do well to heed.

Each species, from the smallest insects to the largest mammals, contributes to the intricate web of life, ensuring the resilience of ecosystems. Humans, with their preference for monoculture and homogenization, often undermine the very diversity that sustains our planet. In the grand mosaic of existence, animals embody the essence of coexistence and symbiosis. The wisdom embedded in their interactions speaks of a harmony that transcends our understanding, a testament to the interconnectedness of all life forms.

Animals understand the deep language of silence—a language lost in the incessant chatter of human minds. Wolves communicate through subtle gestures, elephants through infrasound, and spiders through delicate vibrations on their webs. In this symphony of silence, animals transmit messages more nuanced than the most eloquent human speech. While we drown in the noise of our own creation—be it the constant hum of machines or the relentless clamor of social media—animals navigate their worlds with a profound appreciation for the stillness that eludes us. Perhaps, in our unceasing quest for progress, we have neglected the wisdom hidden in the quiet corners of existence.

Lions rely on synchronized movements of the pride during hunts, cetaceans engage in cooperative hunting, and ants execute intricate maneuvers to protect their colonies. The interdependence and trust inherent in these collective efforts outshine human collaborations. In our pursuit of individualism and self-sufficiency, we often overlook the power that arises from collective commitment. Animals, in their intuitive understanding of unity, remind us that survival is not a solitary act but a fundamental collaboration, where trust forms the foundation of harmonious coexistence.

If ever there were masters of parenthood, they are undoubtedly the creatures sharing our planet. From the attentive care of elephant mothers to the protective instincts of wolves, the animal kingdom exemplifies a mastery of parenting that far exceeds our own. While human parents sign their children up for a series of extracurricular activities, animals impart essential life skills for survival, ensuring the continuity of their species with a natural mastery that renders our parenting manuals ridiculous and inadequate. We have the longest period of dependency, more than any other species. Our parenting as humans involves a very complex process, part of which involves conditioning the child for the difficulties I've mentioned here detaching from our natural state to develop self-awareness and language. We also have our role to play. It's just that most of us aren't doing it!

The lessons from these parents go beyond mere survival—they speak to the essence of nurturing, guiding, and fostering a deep connection with the world. Perhaps, in our rush toward progress, we have overlooked the profound art embedded in the simplicity of raising the next generation.

Even in old age, who would have thought that the animal

kingdom, with its silent wisdom, could teach us something about gracefully accepting the inevitable journey towards gray hair, creaky joints, and the exciting world of bifocal glasses? Let's take a moment to appreciate the great elephants, regal creatures that wander the savannas with the grace of ballet dancers in their senior years. They've mastered the art of aging like a science, leading their herds with an air of sophistication that can only be acquired after decades on this planet. Forget about those aching knees and back pains; elephants have their wrinkles, strips of wisdom etched into their leathery skin, each wrinkle a map of memories that makes our aging process look like a kindergartner's finger-painted picture. And then there are the turtles, the Zen masters of slow and steady. While our society races at the speed of light, these creatures are out there, moving at a pace that makes a sparrow look like an Olympic sprinter. What's their secret? Patience. A virtue that seems to have slipped away with our youth, leaving us with a society too busy seeking the next big thing to appreciate the beauty of aging at a turtle's pace. Maybe we should all trade our running shoes for a good pair of turtle shells and learn to appreciate the art of going slow.

Now, let's talk about whales, the great storytellers of the ocean. Their wrinkles are like the pages of an ancient watery tome, every line telling a story of epic adventures in the deep blue. It's as if they have an entire library etched into their skin, while we struggle to remember where we left our reading glasses. Whales have been there, done that, and worn the worn-out t-shirt of oceanic wisdom, making our midlife crises feel like a tiny hiccup in the symphony of life. Meanwhile, in the human world, we often treat aging as if it's a bad case of the flu—something to avoid at all costs. Our societies have this tendency to relegate the elderly to the margins, as if they're an outdated decorative piece.

We forget that these are the individuals who've faced storms, climbed mountains, and probably have a treasure trove of embarrassing stories about us when we were young. But hey, who needs the wisdom of years when we can chase the elusive elixir of eternal youth? Yes, animals are the true sages of aging with grace, showing us that wrinkles are not just folds in the skin but badges of honor in the game of life. So, the next time someone tells you that aging is a decline, remember the turtle, the elephant, and the whale—the true influencers of the wisdom of aging. Maybe, just maybe, it's time for us to take a page from their book, or should we say, from their wrinkled pages of deep wisdom?

Animals, in their modest brilliance, embody the alchemy of simplicity. Their intuitive understanding of the elements and their seamless ability to adapt to the environment—everything unfolds with an elegance that challenges the complexities of our human existence. As we delve deeper into the fabric of the animal kingdom, we encounter a multitude of species, each with its unique role and contribution to the intricate web of life. Human societies, on the other hand, often find themselves ensnared in the labyrinth of ambition, competition, and technological wonders. In our quest for knowledge, progress, and innovation, we have built elaborate labyrinths of thought and action. The irony is that, despite our intellectual achievements, we often overlook the profound lessons that nature, in its simplicity, is eager to teach us.

Observing the animals around us, it becomes increasingly clear that true genius resides in simplicity. Just look at the remarkable ability of birds to navigate long distances using a magnetic compass, or the instinctive brilliance of a spider creating an intricate web to catch its prey. These acts, seemingly effortless, reveal a depth of wisdom that transcends the need for complicated strategies. The ancient art of survival, etched into the very essence of these creatures, whispers to us from the heart of simplicity. Perhaps, by freeing ourselves from the layers of self-imposed complexity, we can rediscover the wisdom that has patiently awaited us in the quiet corners of the natural world. Animals, in their non-verbal language, invite us to embrace a more harmonious relationship with the world around us. They urge us to simplify, to return to the basics, and to appreciate the beauty of the inseparable.

The alchemy of simplicity is not an invitation to abandon progress or relinquish the advancements that define our civilization. Rather, it serves as a gentle reminder to discern the essential from the superfluous, to find elegance in simplicity without compromising the depth of our understanding. The natural world, with its intricate yet simple designs, presents a canvas on which the art of living with grace is painted by creatures large and small. As we navigate the complex landscapes of our lives, we draw inspiration from the animals that share this planet with us. The unpretentious brilliance of a cat chasing its prey or the communal harmony of a group of odontocetes swimming in unison: these are not just glimpses into the animal kingdom but mirrors reflecting the timeless wisdom of simplicity.

At the heart of every human being lies a primal desire: the call of nature echoing through the ages. It is an invitation to reconnect with the untamed aspects of our existence, to rediscover the instincts that have long been suppressed beneath the veils of civilization. Animals, in their unfiltered existence, embody this call, urging us to step away from the confines of our concrete jungles and embrace the wild nature within us. As we navigate the complexities of our modern lives, let us listen to the call of nature. Let us open ourselves to the rustling of the leaves, the rippling of streams, and the primordial rhythm that beats in the heart of the natural world. In doing so, we may discover that the supposed "stupidity" of animals is merely a reflection of our own disconnected perception—a perception that, when challenged, opens the door to a world of deep wisdom, simplicity, and untamed beauty.

In the delicate choreography of climate modification, animals also reveal the art of living sustainably. Unlike human activities that often exploit resources without considering the long-term consequences, animals have perfected the balance between giving and receiving. The polar bear, symbol of the Arctic, relies on sea ice to hunt seals, understanding the interconnectedness of its existence with the frozen landscape. Humans, on the other hand, are confronted with the consequences of their ravenous consumption, depleting resources at an alarming rate. The animal kingdom, through its climatic dance, invites us to reconsider our relationship with the environment. Can we learn to take only what is necessary, to move in harmony with the natural world rather than imposing our will upon it?

Once again, animals invite us to shed the pride of our technological prowess and embrace the simplicity of living in tune with the environment: each step is a testament to our ability to adapt, harmonize, and thrive on this shared stage called Earth. Animals, in their silent symbiosis, exemplify the art of adapting not only to the environment but within a web of interconnected relationships. Consider the oxpecker bird, which cleans parasites from the back of a rhinoceros.

Perhaps, by embracing a more collaborative stance, we can find solutions that go beyond mere survival to create a symphony of sustainability. The world beneath the ocean's surface is a testament to the symphony of symbiosis. Clownfish find refuge in the protective tentacles of sea anemones, forming a mutualistic alliance that benefits both partners. The clownfish gain protection from predators, while the anemones receive care and cleaning from their colorful tenants. It is a partnership forged not through contracts and negotiations, but through the unspoken language of survival. Humanity, in its quest for dominance and progress, has often overlooked the importance of symbiotic relationships. The animal kingdom, however, is a vivid example of the art of collaboration. Whether it's bees pollinating flowers or bacteria aiding digestion in the stomachs of ruminants, animals demonstrate an innate understanding that survival is not a solitary endeavor but a collective effort. As we face environmental challenges and the consequences of our actions, animals invite us to reassess our relationships, not just with them but with the planet as a whole. Can we cultivate partnerships that go beyond resource exploitation to create harmonious coexistence?

We also have much to learn from mycorrhizal fungi and plants. Beneath the earth, while fungi assist plants in nutrient absorption, the plants provide sugars in return. This underground collaboration, hidden from the human eye, is a reminder that the deepest interactions often occur beyond the spotlight of our awareness. It reminds us that we are not the sole protagonists but merely actors in a vast ensemble. These life lessons invite us to transcend our egocentric approach and embrace a more holistic understanding of our place within the intricate web of life.

Animals seamlessly integrate into their habitats to such an extent that it escapes the human tendency to shape the environment to satisfy individual desires. The intricate architecture of an anthill, the coexistence of plants and insects in the rainforest, and the subtle mimicry of certain species all speak of a wisdom born from adaptation to the environment rather than the imposition of one's will upon it. Humans, with their concrete jungles and sprawling developments, often disturb the delicate balance of ecosystems.

The fabric of evolutionary wisdom tells a story of survival and acclimatization spanning millions of years. Animals, living witnesses to the persistence born from the crucible of natural selection, serve as extraordinary embodiments of life's triumph over environmental challenges. In our relentless pursuit of comfort and convenience, we inadvertently drift away from the unyielding forces of natural selection that have shaped the various inhabitants of the animal kingdom. The adaptability etched into their very DNA is a clear reminder that tenacity is not a virtue cultivated in the ease of predictability but is forged in the relentlessness of environmental adversity. While we may be sheltered by the walls of our modern habitats, animals continue to navigate a world that demands constant adaptability.

Consider extremophiles, those microscopic miracles that thrive in the harshest environments on Earth. From the searing depths of hydrothermal vents to the icy expanses of polar regions, these resilient organisms demonstrate an unparalleled ability to adapt and flourish where many would falter. Their existence challenges our notions of habitability and fortitude, urging us to broaden our understanding of what it means to endure and thrive.

Similarly, the humble cockroach, often vilified for its presence in human spaces, possesses an unflappability that transcends the bounds of our prejudices. With the ability to withstand radiation levels fatal to most living beings, the cockroach stands as a testament to life's endurance in the face of catastrophic challenges. Perhaps, amidst our technological marvels and creature comforts, we can find inspiration in the cockroach's modest ability to persist against all odds.

Taken together, the supposed "inferiority" of animals is a mistaken judgment rooted in human arrogance. As we navigate the complexities of our existence, let us not forget the lessons imprinted in the simplicity of the animal kingdom. True brilliance lies in their ability to adapt seamlessly, to thrive in harmony with the environment, and to embody a wisdom that transcends the limits of our intellectual pursuits. So, while we marvel at our technological successes and intelligence, let us not overlook the quiet revolution unfolding in the natural world. Animals may hold the key to a more sustainable, harmonious, and enlightened existence—one in which the term Animal Stultum becomes nothing more than a reflection of our limited understanding.

24 LOYAL AND FAITHFUL

In human experiences, the exploration of loyalty takes center stage, a chapter often exalted by Homines Stulti as their exclusive domain. Armed with intricate emotions and an evolved intellect, humans claim mastery over a nirvana of loyalty that theoretically sets them apart from the animal kingdom. However, let us not overlook the compelling limbo of loyalty performed by humans, a dance where betrayal is choreographed and trust becomes the sacrificial offering. The circus of human loyalty showcases trapeze stunts of backstabbing and tightrope walks on betrayal, creating a dramatic narrative that often eclipses the genuine connections sought.

In contrast, in the animal kingdom, loyalty is not a negotiable commodity but a steadfast certainty. It is the unconditional love of a dog, whose devotion does not waver based on mood or convenience—a purity of spirit that human behaviour so rarely imitates. A meerkat does not engage in moral deliberations about leaving its clan for a better opportunity. Loyalty is innate, like a survival instinct, an unwritten contract binding animals in their pursuit of sustenance and security. In this rodeo of wild loyalty, there are no competitors reflecting on the depths they can reach before inevitable betrayal. It is a direct commitment, going beyond the complexities of human drama and grounded in primal instincts of survival.

Faithfulness too is often sidelined as yesterday's news, and the allure of infidelity resounds in the corridors of romantic escapades. Intricate dramas unfold with deceptive soliloquies and clandestine meetings, transforming monogamy into an endangered species facing relentless challenges from wandering eyes and questionable decisions.

Even here, animals navigate the labyrinth of faithfulness with a grace that speaks to the simplicity of nature. Wolves mate for life, forming lasting family bonds that withstand the harsh realities of the wild territory. Penguins walk through life flipper in flipper, unwavering in their commitment amidst icy landscapes. Beavers build homes for their families with unparalleled dedication, free of the TV scenarios and love triangles that characterize human relationships. Their commitment is direct, written in the language of nature without the need for complex narratives.

As we delve deeper into the exploration of human loyalty, nuances emerge, revealing the complexities of a species that prides itself on emotional depth and intellectual acumen. The human capacity for tolerance, loyalty, and faithfulness is not confined solely to theatrical performances and circuses.

Within the complex web of family bonds, most humans show a unique blend of loyalty, tolerance, and faithfulness. Parents often invest unconditional dedication in raising their children, demonstrating a commitment that goes beyond the theatrical performances seen in social interactions. Siblings, bound by blood and shared experiences, navigate the landscape of loyalty with a blend of camaraderie and inevitable conflict, offering a microcosm of human relationships.

Among friends, the dynamics of loyalty and tolerance take on a different shade. The limbo of loyalty may be less pronounced, but the subtleties of human interactions come to the forefront. Friends tolerate quirks, face challenges of loyalty, and grapple with the complexities of faithfulness. In this context, human capacity for genuine connection and emotional depth stands out, offering a contrast to the more performative aspects seen in broader social contexts.

Broadening our view to the larger framework of social structures, loyalty manifests in different forms. Citizens pledge loyalty to their nations, showing a collective loyalty that transcends individual relationships. Institutions, whether political, religious, or social, thrive on the loyalty of their members, shaping the fabric of societies. Tolerance for different perspectives and loyalty to shared values become integral components of the narrative of social loyalty.

However, in the realm of Homo Stultus, the line between

sincerity and artifice blurs. Tolerating quirks may be a genuine acceptance for some, while for others, it is a mask worn to conform to society. Loyalty can be an unconditional commitment to an individual, or a facade for another seeking personal gain. Faithfulness can be the cornerstone of some relationships and a victim of human flaws in others.

In contemplating the labyrinth of human loyalty, one cannot help but cast a glance at the simplicity and authenticity of the manual of the animal kingdom. While humans struggle with the complexities of tolerance, the limbo of loyalty, and the elusive nature of faithfulness, animals silently navigate a world where commitment is instinctual and not adorned with the complexities of human drama.

The animal kingdom imparts lessons in direct communication, honesty, and an unshakable commitment to survival. The absence of hypocrisy in their interactions stands in stark contrast to the intricate complexities of human social dynamics.

As we take a look at the "Manual of the Animal Kingdom", it becomes useful to explore how these lessons might be applied to the complexities of human relationships, loyalty, and interconnectedness.

It is an invitation to learn from the wisdom of the animal kingdom. By embracing honest communication, recognizing loyalty as a survival instinct, simplifying commitment, prioritizing genuine connections, reconnecting with instinctual wisdom, and learning from loyalty in different contexts, individuals can discover that the deepest loyalty does not require a script or applause: it thrives in the authenticity of shared existence.

25 THE WILD SLEEP

Night falls, darkness wraps the earth, and living beings prepare for their rite of passage: sleep. However, there is an abyssal difference between how we humans face this vital need and how animals manage it with mastery. Let's discover why the animal kingdom easily wins this sleep competition!

Humans tend to make questionable food choices before going to bed. We indulge in hearty dinners and late-night snacks while hopping from one TV series to another. Our pre-sleep food is a mix of carbohydrates, fats, and sugars, causing our stomachs to compete with our brains for the title of "Who Can Keep Poor Sleep Awake." On the other hand, wild animals prepare for rest with a more sober diet. No late-night meals for them, thanks. At most, a bite of fresh grass or a recently caught prey. Animals know that a light stomach means a better sleep.

Moreover, humans have developed all sorts of theories about the best way to sleep. We've read everywhere that sleeping on our back is the best way to maintain proper posture and prevent wrinkles. But who told humans they can't have quality sleep unless they toss and turn like they're in a salad? Wild animals, on the other hand, don't need to search for YouTube tutorials on "how to sleep correctly." They stretch out on a branch, curl up in a den, or simply lie in the sun. Simple and effective.

While we humans seek the comfort of our beds, we often neglect the importance of creating a sleep-friendly environment. "But who needs total darkness to sleep?" we ask, while turning on the bedside lamp and leaving our phone and TV on. Our brains, desperately trying to switch off, must face an epic battle against all these sources of light and sound. Meanwhile, wild animals are experts at creating the perfect sleep environment. A cozy nest, a well-protected den, or simply a quiet corner in the tall grass. No phone ringing, no blinding lights, just the serenity of nature.

And speaking of serenity, it's important to consider how animals, unlike humans, respect their natural circadian rhythm. While we struggle against our biological clocks, trying to adapt to a rhythm imposed by modern society, animals surrender to the natural flow of light and darkness. Lions, tigers, and bears, for example, wake up and sleep in harmony with nature's cycles. They don't need loud alarms or coffee to start their day; they simply follow the signals of their bodies. Humans, on the contrary, often ignore these signals, disrupting their sleep schedules with negative consequences for their health.

Another fascinating aspect of animal sleep is the variety of postures and places they choose to sleep. Take, for example, bats, which sleep hanging upside down, or flamingos, which manage to sleep on one leg. Whales and cetaceans, meanwhile, adopt a completely different strategy: they sleep with half of their brain at a time, allowing them to stay in motion and breathe. This unique customization enables them to avoid predators and maintain vital functions even during rest. Humans, on the other hand, are limited to a much narrower range of positions and places where we can sleep comfortably.

Let's also consider the social aspect of sleep in the animal kingdom. Many animals find safety and warmth by sleeping in groups. Emperor penguins huddle together to protect themselves from the Antarctic cold, while wolves curl up in packs to retain body heat. Humans, although sometimes sharing a bed with a partner or a pet, rarely experience sleep as a collective activity with the same intensity and meaning it holds for many animals.

Protection techniques during sleep are another area where animals excel. Some species, such as horses and elephants, sleep standing up, ready to spring into action at any sign of danger. Other animals, like lemurs, hide among tree branches or in deep crevices to avoid predators. Humans, despite closed doors and security systems, often cannot match the ability of animals to adapt and protect themselves during rest.

In terms of sleep duration, animal habits are equally varied. While some species, like lions, can sleep up to 20 hours a day, others, like giraffes, settle for just a few minutes of sleep at a time.

This variation is not only fascinating but also indicative of how each species has evolved specific sleep strategies suited to their ecological and biological needs. Humans, by contrast, tend to standardize our sleep needs, often overlooking how much these can vary individually.

Ultimately, while we humans struggle to fall asleep, wild animals sleep peacefully, without worries and without the need for apps to monitor their sleep quality. Perhaps we should take a lesson from them and embrace our wild side. And who knows, maybe we too could enjoy a night of deep, rejuvenating sleep, without having to fight against our own modern lifestyle. By integrating these lessons into our lives, we might not only improve our sleep quality but also rediscover a deeper connection with nature, which has always supported us.

CONCLUSION

At the end of this journey through the intricate relationships between humans and animals, we find ourselves with a deeper awareness of the complexities that characterize our relationship with the rest of the animal kingdom. Through sarcasm and reflection, we have explored topics ranging from boredom to loyalty, from dental diversity to habitat destruction, offering new perspectives on issues often overlooked or trivialized.

In this brief essay, we have embraced an unusual approach: that of siding with the animals. Through this lens, we have sought to challenge our preconceived beliefs about ecology, ethology, and our own place in the natural world.

However, this book is not merely an exercise in sarcasm or intellectual provocation. It is also an invitation to greater awareness and sensitivity toward the other forms of life with whom we share the planet. Each chapter has been designed to stimulate reflection and promote a deeper understanding of the dynamics that shape the interactions between humans and animals.

In pursuing a better understanding of the differences and similarities between humans and animals, we hope this book has sparked greater empathy and a heightened awareness of our responsibilities toward the natural world. Whether it is reconsidering our role in the ecosystem or reflecting on the ethical implications of our actions, we hope that Homo Stultus continues to stimulate discussions and reflections even after its reading.

ABOUT THE AUTHOR

Matteo Preabianca is a linguist, educator, and translator whose work reflects over twenty years of travel across different countries. Fluent in Italian and English, he expresses his dual cultural identity through a diverse portfolio that includes poetry, experimental music, and novels. A vegan since the age of 14, he has a deep interest in animal rights and their suffering, themes that influence both his personal and professional journey. He earned a PhD with honors in Education and Socio-educational Intervention and currently resides in Scotland, where he teaches English to Ukrainian refugees.

DISCUSSION GUIDE FOR SMALL GROUPS, CLASSES, OR INDIVIDUAL STUDY

1. In the introduction, the author adopts a sarcastic tone to explore the relationship between humans and animals. How does this tone encourage readers to question commonly held beliefs about human superiority?
2. In the chapter "Orientation," the author describes the first steps in understanding animal behaviour. How do these initial explorations set the stage for the book's deeper insights?
3. The chapter titled "The Bonobo Experiment" highlights a fascinating comparison between human and animal behaviour. What does this experiment reveal about the similarities humans share with other species?
4. The book discusses boredom as a concept. What does this exploration of boredom reveal about both human and animal life, and how does it challenge our perception of intelligence?
5. In "How Much We Talk!," the book reflects on human communication. How does this contrast with animals' non-verbal communication, and what does it suggest about the complexity of language?
6. "An Animal Doctor!" explores healing in animals. How does this discussion prompt a rethinking of human medicine and the role of nature in health?
7. The chapter "And What About Antibodies?" compares the immune systems of humans and animals. How does this comparison deepen your understanding of survival mechanisms in life?
8. In "How Good Is Arsenic!", the author explores animals thriving in extreme conditions. What lessons can readers take from these survival tactics, and how might they

challenge human limitations?

9. "Let's Save Ourselves: When Nature Gets a Head Start on Humanity" suggests that nature often solves problems before humanity does. How does this chapter prompt reflection on human innovation and our relationship with the natural world?
10. In "Existential Problems," the book discusses the complexity of life. How do the philosophical reflections in this chapter influence your understanding of both human and animal existence?
11. "A Story Without Ghosts" examines human imagination compared to animals' lack of it. How does this difference highlight alternative ways of experiencing existence, free from complex thoughts?
12. In "Dental Diversity," the book explores how animals' teeth reflect their survival needs. How does this diversity illustrate the ways animals adapt to their environments?
13. "Pain Tolerance" examines how different species experience pain. What insights can be gained from animals' pain tolerance, and how might this impact human empathy toward them?
14. The significance of hair and beards in animals is discussed in "Beard and Hair, Thank You!". How do these physical features relate to survival, and what parallels can be drawn to humans?
15. "Masters of Survival" looks at animals thriving in tough environments. What lessons on adaptability and resilience can humans learn from these survival strategies?
16. The "Complicated Trachea" chapter explores the intricate anatomy of certain animals. How does this challenge human-centric views of complexity in the natural world?
17. "Underwater" and "Flight" focus on unique forms of mobility. How do these chapters shape your understanding of freedom and physicality in the animal kingdom?

18. "Shock Wanted" examines stress and survival. How does understanding the animal response to shock influence your perspective on how humans cope with stress and adversity?
19. In "Let's Not Climb," the author discusses the limits of certain abilities. How does this chapter highlight the boundaries of both human and animal capabilities?
20. In the conclusion, the book reflects on the human-animal relationship. What key insights do you think readers are meant to take away from this exploration of the world through an animal-centric lens?

🌿 How to Leave a Review Guide for a ShelteringTree.Earth Book

Thank you for supporting our authors. Reviews help readers discover books that speak to their hearts and spirits. Even a short review makes a meaningful difference.
Here's a simple guide to help you leave a review on Amazon, Goodreads, ShelteringTreeMedia, or any other book review site.

✅ 1. What to Write in a Review

Your review does *not* need to be long or complicated.
A few sentences is enough.
You can share:

- what you enjoyed
- what you learned
- how the book made you feel
- who you think would benefit from it
- your favorite part or takeaway

You do **not** need to summarize the entire book.

✅ 2. How to Leave a Review on Amazon

Step 1: Go to the book's Amazon page
Search for the title or use the link provided by the author.
Step 2: Scroll down to "Customer Reviews"
Click **"Write a customer review."**
Step 3: Choose a star rating
5 stars = excellent
1 star = poor
Step 4: Write your review
A few sentences is perfect.
Step 5: Click Submit
That's it — your review is live.

☑ 3. How to Leave a Review on Goodreads

Step 1: Log in to Goodreads

(You can create a free account if you don't have one.)

Step 2: Search for the book

Click on the correct edition.

Step 3: Click the stars to rate the book

This alone counts as a review.

Step 4: Click "Write a Review"

Add your thoughts and click **Save**.

☑ 4. How to Leave a Review on ShelteringTreeMedia.com

Step 1: Go to the ARC tab on ShelteringTreeMedia.com

Step 2: Scroll down to the Review box

Step 3: Fill in the form

Step 4: Click SUBMIT

ARC Team members will be sent a coupon once the review is posted under the book review blog.

☑ 5. Tips for Helpful Reviews

- Be honest
- Be kind
- Be specific
- Keep spoilers minimal
- Share how the book impacted you

☑ 6. Why Reviews Matter

Reviews help:

- authors reach new readers
- bookstores and libraries decide what to carry
- online algorithms recommend the book
- readers discover books that nourish their spirit

Your voice truly makes a difference.

SHELTERING TREE
Earth Publishing
ShelteringTreeMedia.com

www.ingramcontent.com/pod-product-compliance
Lightning Source LLC
Chambersburg PA
CBHW071446230926
51846CB00001B/361
* 9 7 8 1 9 6 3 2 7 2 9 8 7 *